Managi[illegible] Construction Logistics

Gary Sullivan
Wilson James Ltd

Stephen Barthorpe
MITIE Group Plc

Stephen Robbins
Laing O'Rourke

WILEY-BLACKWELL
A John Wiley & Sons, Ltd., Publication

This edition first published 2010

Blackwell Publishing was acquired by John Wiley & Sons in February 2007. Blackwell's publishing programme has been merged with Wiley's global Scientific, Technical, and Medical business to form Wiley-Blackwell.

Registered office
John Wiley & Sons Ltd, The Atrium, Southern Gate, Chichester, West Sussex, PO19 8SQ, United Kingdom

Editorial offices
9600 Garsington Road, Oxford, OX4 2DQ, United Kingdom
2121 State Avenue, Ames, Iowa 50014-8300, USA

For details of our global editorial offices, for customer services and for information about how to apply for permission to reuse the copyright material in this book please see our website at www.wiley.com/wiley-blackwell.

Library of Congress Cataloging-in-Publication Data

Sullivan, Gary, 1959–
Managing construction logistics / Gary Sullivan, Stephen Barthorpe, Stephen Robbins.
p. cm.
Includes bibliographical references and index.
ISBN 978-1-4051-5124-5 (pbk. : alk. paper) 1. Building–Superintendence–Data processing. 2. Construction industry–Management–Data processing. 3. Business logistics–Data processing. 4. Production scheduling–Data processing. I. Barthorpe, Stephen. II. Robbins, Stephen, 1979– III. Title.
TH437.S83 2010
624.068′7–dc22

2009048117

A catalogue record for this book is available from the British Library.

Set in 10.5 on 12.5 pt Sabon by Toppan Best-set Premedia Limited
Printed and bound in Singapore by Fabulous Printers Pte Ltd

1 2010

Contents

Foreword

How do you get a supply line to work efficiently? All industries have to tackle that question in order to succeed, but it has, to date, proved a particularly thorny issue for the construction industry.

There are many complex and interrelated reasons for this. The industry has a low entry requirement in terms of capital, and many sizable firms have grown incrementally from the humblest beginnings. Despite the capital required to fund a major project, the industry is not perceived as capital-intensive, and thus fails to get the business attention and investment that is accorded to many comparable sectors. Furthermore, the fragmentation of the industry creates a silo mentality, where everyone is focused on their own area of expertise rather than investing their resources into achieving excellence in the project as a whole.

For these reasons, a kind of amateurishness persists in the construction of many projects. If you walk past the average building site, the wastage is immediately apparent. New materials have been damaged before they could ever be used and lie abandoned in skips. Even those materials that are used are often dented or scratched en route because of poor management processes. These damaged items can be repaired, but they will never again be pristine. In effect, the client is paying for a second-hand building.

These issues are certainly not new. When I began collaborating with Gary Sullivan in the late 1980s, I was increasingly keen, as a client, to improve the way building sites were managed. Gary was providing site security at Bovis Construction and was starting to become interested in improving the efficiency of both this function and other prelim services on site.

I suggested that we asked the security staff to be helpful and welcome people, and to start making a proactive contribution to site organisation. I wanted them to be seen as people who helped with the core business of the project rather than as an obstructive presence that impeded progress. Gary took on the challenge of improving their effectiveness and providing them with additional skills. Before long, we had started to look at real logistics in earnest, exploring ideas and solutions from other industries to see how they could be adapted for construction. He ultimately developed this to the point where we were finding better ways of managing the infrastructure of entire projects, including vastly improved handling of both materials and waste.

Since then, Gary has become a leader in translating ideas from other sectors to make them work in construction. His passion has driven many important developments, and he's invested immense amounts of time and money to get new solutions to work. Some enlightened clients have adopted these techniques, and are reaping the benefits.

However, twenty years after we realised what a professional approach to logistics could achieve, there are still vast swathes of the industry that are failing to grasp this opportunity to improve their performance. Too often, logistics contractors become vulnerable to budget cuts, because, despite their contribution to productivity, they're not perceived as core to the project. The decision-makers involved simply don't realise that this function can often determine the difference between the on-time delivery of a gleaming new building and one where damage, delay and claims are the predominant feature.

Since the adoption of professional logistics in construction has been both piecemeal and slow, the casual observer might be tempted to conclude that widespread change will never occur in this most traditional of industries. That conclusion, I firmly believe, is wrong.

Perhaps the best analogy that I can draw is with the shipbuilding industry. Radical change to delivery methods arrived in the form of containers, rendering traditional practices redundant. Still, when the American entrepreneur Malcolm McLean introduced them in 1956, 'no one anticipated that his initiative, which traditional ocean carriers considered utter folly, would transform the packing and carrying of general cargoes worldwide'[1]. The technology had already been in existence for several decades, although it had not been widely adopted. Suddenly, a combination of circumstances relating to both market forces and labour relations triggered its adoption by a critical mass of shipping operators. Once this happened, docks that were unable to adapt would not survive.

The parallels are striking. Until a critical mass of businesses adopted it, the container had been ignored for decades by an industry content to use technologies that would have not have been unfamiliar to the ancient world. Once the critical mass was reached, however, no one could ignore it or compete against it. The industry changed forever.

In just the same way, the construction industry will reach a critical mass in its adoption of professional logistics, and I believe it will do so

[1] Donovan A (1999) Longshoremen and mechanisation: A tale of two cities. *Journal of Maritime Research* December issue.

sooner rather than later. When this momentum really begins to build, the step-change will dwarf all the developments to date.

The multi-skilled logistics and security team is now a proven benefit. So far, we have been able to achieve a certain amount of efficiency through the use of construction consolidation centres. Still, at the time of writing, there is only one operating in central London. In the future, we, as an industry, will need five or six centres around London, to which all materials coming into the capital will be delivered. The revolutionary effect on the road system alone can only be imagined. Instead of half-empty juggernauts struggling through narrow and congested streets, smaller and more nimble lorries will deliver consolidated work packs across the city. After decades of development, the consolidation centre, just like the container ship, will suddenly become an overnight success. No organisation that fails to buy into this newly efficient and environmentally friendly incarnation of the industry will survive. Its wastage levels will be too high, in comparison to its peers, for it to continue.

I believe that constructors ignore this book at their peril. These techniques will soon be essential to every construction business. The only question for industry decision-makers is to decide whether they want to get ahead of the curve by changing now or wait until a streamlined and competitive industry forces them to adapt just to remain viable. As a client, I know what I'd recommend.

Peter Rogers CBE
Chairman of the UK Green Building Council

Preface

Readers of this book will not have failed to notice the proliferation of transport company vehicles that operate on our motorways that feature the word 'logistics' in their name and logo. What was, for decades, known appropriately as 'Smith & Son Transport' has inexplicably evolved into 'Smith & Son Logistics' – without any apparent diversification of the services offered. 'Logistics' has inaccurately become synonymous with 'transport' and as such misrepresents and significantly undervalues the scope of true logistics – especially when applied to the construction industry. The following pages of this book will demonstrate that logistics is multi-faceted, with many interdependent functions and, if managed properly, will make a significant contribution to the successful completion of construction projects.

Managing Construction Logistics is aimed at everyone interested in making construction site logistics work more smoothly. Construction delivers complex bespoke projects, often in challenging and constrained environments, so it has much need of professional logistics solutions. Despite this, logisticians have paid but sparse attention, to date, to the construction industry, and very little has been written about the best way of organising and supplying a project. This both reflects and perpetuates the lack of collaboration between the professional communities of logistics and construction.

This book is perhaps the first sustained attempt to remedy this situation, having been co-authored by professionals from both disciplines. Gary Sullivan, a professional logistician who has specialised in construction for twenty years, has written Section Two, which gives insights into the best way of managing site logistics from a practitioner's perspective. Stephen Barthorpe and Stephen Robbins, both construction professionals with strong academic credentials, were jointly responsible for the contextualisation of logistics in Section One, and for the analysis of future trends offered in Section Three. It is hoped that this juxtaposition of the practical and theoretical will make this work of use to both practitioners and students.

Inspiration for this book evolved following the authors' collaboration on a research project investigating 'The Potential for Transferring the Heathrow Consolidation Centre Methodology to other areas of the Construction Industry'. The project was jointly funded by the Department of Trade and Industry, Wilson James, Mace and BAA and culminated in the publication of a final report, *Consolidating Construction Materials: Transferring the Methodology*, in 2004.

Some readers may notice that there is no health and safety chapter in the following pages. That is because the authors firmly believe that a commitment to health and safety should be part of every activity on site. As we have tried to make clear, good logistics can have a hugely beneficial effect on the safety performance of a project, as well as minimise the risk of long-term injuries through poor manual-handling practices. When functioning at their best, logistical solutions can even ameliorate the stress that a major project can engender. Therefore, health and safety has not been tucked away in a discrete chapter. Awareness of the need to keep everyone on site healthy and injury-free is at the heart of this text, just as it should be at the heart of any construction project.

About the Book

Section 1: Contextualising Logistics for Construction

This section introduces the concept of logistics and provides definitions and an overview of its origins and development. From its original military applications, where battles were often won or lost depending on how well the logistical support was implemented, to the more recent, universally applied supply chain resource applications, logistics has developed and become generally accepted as an integral facet of modern business practice, particularly in the retail and manufacturing industries.

Logistics involves the strategic and cost-effective storage, handling, transportation and distribution of resources. It is an essential process that supports and enables the primary business activity (e.g. a construction project) to be accomplished.

Traditionally, insufficient attention has been given to logistics. Seminal reports and reviews by Latham (1994), Egan (1998), Bourn (2001) and the Strategic Forum for Construction (2002) have drawn attention to the UK construction industry's wasteful procedures, poor productivity performance and inefficiencies. Improved logistics is highlighted among their numerous recommendations for improvement. The Strategic Forum for Construction Logistics report *Improving Construction Logistics* in 2005 identified many reasons for why the UK construction industry has not universally adopted dedicated logistics, notably the fragmentation of the industry and the one-off, bespoke nature of construction projects that militate against implementing long-term change.

This section also introduces the concept of a dedicated approach to logistics and the emergence of a specialised logistics management approach for the construction industry. Effective logistics management is crucial to the success of modern businesses, especially businesses or projects that rely on extended supplier networks and just-in-time deliveries. The global sourcing of products and services has introduced additional complications, exacerbating the logistical challenges. Dedicated construction logistics has developed into an essential support service and has led to the emergence of dedicated, specialist logistics service providers like Wilson James Ltd.

Section 2: Construction Logistics in Practice

This section describes more fully the diverse range of support service applications of construction-specific dedicated logistics management. It is written from a professional practitioner's viewpoint and as such provides practical guidance on implementing successful logistics on complex and challenging construction projects. The author of this section is uniquely qualified and offers a candid, pragmatic and often acerbic analysis of working in an industry that could justifiably be described as dysfunctional.

Using a dedicated logistics management approach either by employing a specialist logistics contractor or by utilising a directly employed logistics team enables the principal contractor to concentrate on its primary construction business and have the non-core, albeit essential, construction logistical issues managed by specialists more effectively.

The range of site-based logistical support is varied and includes essential functions such as security, materials handling and delivery, waste management, traffic management, temporary works, welfare, emergency services liaison, catering and housekeeping. Although these functions could be considered non-core construction activities, unless they are effectively planned and successfully implemented the overall project performance will suffer.

Section 3: The Future of Construction Logistics

This section introduces the consolidation centre concept and methodology (CCM) and provides a compelling argument for applying it to the construction industry based on its successful use over many years in the retail and manufacturing industries.

The first example of a dedicated logistics facility for a major construction project in the UK is the consolidation centre (CC) at London's Heathrow Airport. This innovative facility was the subject of a Department of Trade and Industry (2004) research project undertaken by the authors to investigate the potential for transferring the CCM to other areas of the construction industry.

CCs are essentially storage facilities that hold materials or equipment for a limited period prior to delivering them to the point of use on the construction site, on a just-in-time basis. Unlike warehouses that may store things for long periods, CCs intentionally provide short-term buffer storage to reduce congestion, potential theft or damage and inconvenience when otherwise transported directly to site and stored. The CC ideally suits sites that have restricted access due to their geographical location, restricted access or high levels of security (airports, prisons, law courts, military projects, pharmaceutical establishments etc.).

The logistical, environmental and social benefits of using a CC are compelling, and these are described in detail in this section.

Although three distinct types of CC are proposed in this book, a hybrid version of two or all three could also be considered:

Concealed consolidation centre The concealed consolidation centre derives its name from its intended location as it is concealed within the boundary of a site's perimeter. This is the most basic form of CC and is usually operated by the principal contractor managing the project.

Communal Consolidation Centre The communal consolidation centre is aptly named because of its intended purpose of serving numerous single-client or single-contractor projects, typically no more than about 5 km from each other.

Collaborative Consolidation Centre The collaborative consolidation centre derives its name from the collaborative nature of its shared use between different clients and contractors. It is the largest and most sophisticated type of CC and would be intended to serve multiple sites simultaneously over a wide geographical area and be operated by a specialist logistics contractor.

Section 3 also provides seven case study examples demonstrating how innovative logistics has been successfully used on construction projects in the UK.

The book concludes by proposing a compelling argument for change through implementing a dedicated logistics management approach and by applying CCM to the construction industry.

Stephen Barthorpe

References

Bourn J (2001) *Modernising Construction*. National Audit Office, The Comptroller and Auditor General, London.

Department of Trade and Industry (2004) *Consolidating Construction Materials: Transferring the Methodology: Final Report*. DTI, London.

Egan J (1998) *Rethinking Construction*. The Report of the Construction Task Force to the Deputy Prime Minister, John Prescott, on the Scope for Improving the Quality and Efficiency of UK Construction, Department of Environment, Transport and Regions. HMSO, London.

Latham M (1994) *Constructing the Team*. HMSO, London.

Strategic Forum (2002) *Accelerating Change*. Rethinking Construction, London.

Strategic Forum for Construction Logistics Group (2005) *Improving Construction Logistics*. Construction Products Association, London.

About the Authors

Gary Sullivan MBA is co-founder of Wilson James Ltd, which provides a range of support services to industry, commerce and government, some of which include logistics, security, construction consolidation centres consultancy to many blue-chip clients across the UK. Wilson James Ltd employs over 2,000 people nationwide.

After leaving the armed forces, Gary worked freelance in a variety of industries, including petrochemical, pharmaceutical, event management and construction.

In the early 1990s, he joined the United Nations High Commission for Refugees as part of the team that set up humanitarian aid and logistics in war-torn Yugoslavia.

Gary is Chairman of the Thames Gateway South Essex Partnership, a Board member of Renaissance Southend and Chairman of the Essex Strategic Board for Legacy for the Olympic Games.

Stephen Barthorpe BSc, MSc is the Corporate Responsibility Manager for MITIE Group PLC, a major strategic outsourcing and asset management company. He is responsible for implementing the Group's corporate responsibility strategy throughout the UK.

Prior to joining MITIE in 2005, Stephen was a principal lecturer at the University of Glamorgan for 13 years, researching, publishing and teaching strategic management topics, including corporate social responsibility, corporate culture and environmental issues. Stephen's construction-related interests focused on construction project management, with a particular interest in logistics, and he was the first academic to embed this important topic into the undergraduate construction management curriculum in 1993.

Stephen was a founding director and monitor of the Considerate Constructors Scheme, and assessed constructors' corporate responsibility performance against the Code of Considerate Practice on over 400 construction projects.

Before his academic career, Stephen worked in various senior project and contract management roles in the construction industry for 14 years in the UK and Zambia.

Stephen is a Chartered Builder, Chartered Building Engineer and Chartered Environmentalist.

Stephen Robbins BSc (Hons) MPhil graduated from the University of Glamorgan in 2003. He then worked as a researcher for Wilson James Ltd, authoring a commissioned report, *Consolidating Construction Materials: Transferring the Methodology*, for the Department of Trade and Industry. This was published in October 2004.

Stephen then returned to the University of Glamorgan, where he completed an MPhil thesis entitled *Revolutionising Construction Logistics by Implementing the Consolidation Centre Methodology*.

Since 2005, Stephen has been employed by Laing O'Rourke, where, as off-site production manager, he has had the opportunity to use his knowledge in the development and implementation of logistics strategies for a number of challenging projects, both at tender and at contract stage.

As the emphasis on corporate social responsibility within the construction industry grows, Stephen is becoming particularly interested in the social benefits of implementing logistics management on sites that are constrained by their location, focusing on the need to reduce the negative impact of materials delivery on neighbouring stakeholders working or living near the site.

Dedication and Acknowledgements

To Paul, Devon, Sean, Ali and Dobbo

Acknowledgements

It's fair to say that the book wouldn't have been written, and construction logistics would not be as advanced, if I'd not received a huge range of support from project teams, principal contractors and clients. Many people have helped, but it would be somewhat ungracious not to acknowledge a few people by name. In no particular order, I'd like to thank Mark Reynolds and Ian Eggers for their enthusiasm for logistics, and for teaching me about construction and its politics. I would also like to thank Steve Pycroft for his support in our early days, Dr John Connaughton for his rigour in exploring the benefits of consolidation centres and construction logistics, Paul Sims for his encouragement and support (but most of all because he gets it) and of course my personal holy trinity of Peter, Paul and Steve at Stanhope. Last but not least, this book would not have been written without Ruth Bonner and Caroline Collier.

Gary Sullivan

Glossary of Abbreviations

ASN	advance shipment notice
CC	consolidation centre
CCM	consolidation centre methodology
CCS	Considerate Constructors Scheme
CDM	Construction (Design and Management) Regulations 2007
CMA	construction management agreement
COSHH	control of substances hazardous to health
CPA	Construction Products Association
CSCS	(Construction Skills Certification Scheme) card
CSR, aka CR	corporate social responsibility
HCC	Heathrow Consolidation Centre
ICT	information and communication technology
IFRIT	infrared identification tagging
JIT	just-in-time (delivery)
LOLER	Lifting Operations and Lifting Equipment Regulations 1998
MDUs	mains distribution units
MEWPs	mobile elevating work platforms
MHE	manual-handling equipment
PAT	portable appliance testing
PIR	passive infrared lighting
PPE	personal protective equipment
RFID	radio-frequency identification
SCM	supply chain management
SIA	Security Industry Authority
SMEs	small- and medium-size enterprises
SWMP	site waste management plan
UIN	unique identification number
VSM	value stream mapping
WMS	warehouse management systems
WRAP	Waste and Resources Action Programme

Introduction

The purpose of this introduction is to give readers some insight into the philosophy – and experiences – behind this book.

I first learnt the importance of good logistics in the army. This was then reinforced by my civilian career, which took me in a variety of directions connected to the security industry. Security and logistics go hand in hand: where moving high-risk shipments or people are concerned, meticulous route planning and organisation are essential to successful protection.

In the late 1980s, I was asked to provide security for a Bovis Construction Ltd project in a sensitive location – extra precautions were needed due to its proximity to the Old Bailey law court. That project was Ludgate railway works.

The security was relatively straightforward to set up, and I found myself with time to look round the site. I was fascinated by the skill and dexterity of the trades, and it was clear that they were going to deliver spectacular buildings. Still, some things about the site bemused me. Why was there no methodical approach to logistics? None of the organisational elements appeared to be joined up. Wouldn't it be easier to do things differently? Scores of ideas presented themselves. I was starting to get interested in why things were done the way they were.

Next door, filming for a television programme called *Rumpole of the Bailey* had been taking place. Like a lot of dramas, it was shot out of sequence. Watching it some time later, I realised that the backdrop to the action changed from a demolition site to a half-formed building – apparently during the course of a single conversation! Watching that miraculous transformation over the course of a few minutes really made me interested in the logic behind the sequencing and programming of construction projects.

As time went on, I became more involved with site meetings. Some of my ideas – such as putting loading bays in at the start of a project – I still adhere to today. Other ideas, I quickly learnt, could not be done. I was lucky to be working with an exceptional group of professionals, who took the time to answer my questions. Still, I was frustrated by the difficulty of implementing change, and my security team was finding the sometimes chaotic environment difficult to work in. After a particularly heated difference of opinion with a project manager about whether certain changes were possible, I was almost ready to give up. However, a stranger had been standing behind me, listening to what I was saying. He asked me to write a paper on my ideas. His name was Peter Rogers, and he worked for the client, Stanhope. The colleague I was debating with gave me some pointers and encouraged me to write it, and the paper was duly submitted. Peter called me, and I was invited to a meeting to discuss its contents.

At that time, Stanhope was just finishing Broadgate, a highly innovative project delivered using what was then a very new approach – construction management. They were finding that the trade contractors were very good at doing what they specialised in, but they weren't used to a system where they needed to organise themselves as well. Clearly, a sea-change was required for construction management to work. Peter had quickly realised that they needed a solution to the logistics issue, but at the time there was no firm able to supply it to them. There were transportation specialists, plant suppliers, temporary accommodation and labour gangs, but no one had joined it all up.

Peter explained their cultural and organisational problems. I suggested potential solutions. In return, Peter provided me with the mentoring I needed to get to grips with the way the industry worked. In addition to Peter's help, Ludgate was a project brimming with many of the most promising young managers in construction, as well as a lot of highly respected trade contractors. It was luck rather than good judgement, but I had been given the chance to learn from some of the best in the business. As a believer in training and development, I was also fortunate that Bovis Construction took time to teach me how construction happens.

Ludgate was a hugely successful project, but it was also the birthplace of construction logistics. It would be totally inaccurate to pretend that all the innovations we attempted worked – sometimes it was the proving ground for what not to do – but we learnt a lot. Most importantly, we got enough right to prove that good logistics could make a difference.

Over the life of the project, I had looked at all the operational functions of a site, particularly the prelims and the movement of materials and waste. I'd created efficiencies by training the security guards as banksmen and slingers, realising how much these often marginalised staff wanted to be part of the project. Instead of security guards in pseudo-military uniforms with shiny buttons, they were issued with workwear and taught to multi-task. In short, the logistics and security functions had become a credible, united team with a range of skills. The logistics function was directly employed by Bovis, and Stanhope, as the client, supported the initiative by providing the money to train people to multi-task. I had permission from my line manager to go straight to Stanhope with any issues. Things got done. People started to invent and suggest things, such as a system of bins for organising waste. The experiment had been a success.

During Ludgate, of course, I was still on a huge learning curve when it came to understanding the business of construction. Essentially, my ideas and expectations had been formed in the military. I understood a lot about solving practical problems and achieving results. The cultural and contractual complexities of construction, however, took me a lot longer to understand.

Perhaps the biggest jolt was the end of the project. I was used to battalions whose formation stayed largely the same over time. Successful teams, in the army, were kept together. Nothing prepared me for the moment where, without any fanfare, the project was over and the team was scattered to the four winds.

When the project finished in the early Nineties, recession was tightening its grip on the UK. There were too many people and not enough projects, and logisticians were deemed a luxury that construction didn't need. I had to think about what to do next. Ludgate could have become my first and last construction project, because I was invited to work with the United Nations' High Commission for Refugees to deliver humanitarian aid to Bosnia, at that time a war zone.

It was the first time the UN had attempted to deliver aid in a war zone. My job was to help set up an efficient warehousing and transport operation in a complicated war with three sides, for an organisation that had never delivered in a live war zone before. It was a strange and bloody war. On the one hand, there was a descent into barbarism possibly unequalled in Europe since 1945; on the other, petty Eastern Bloc regulations remained in force. It was an interesting experience, managing time, resources, administration and health and safety in the fog of war without the structure of the army, with all its systems, training and doctrine. And yet, whilst there was little structure, there was plenty of bureaucracy. Chaos, cruelty and compliance coexisted.

It was within this strange environment that I reflected on my time at Ludgate, gaining some crucial insights into the nature of construction. Although many of my UN colleagues and I had a military background, many of the people we worked with were professional humanitarians or from non-governmental organisations, such as charities. Like construction companies, they had to tender for contracts. Doing good was a commercial, cut-throat process. I found myself trying to work with a range of different organisations who, just like trade contractors, were bidding for contracts, competing for funds and falling out. Working for the UN turned out to have more in common with working in construction than with serving in the army. In the army, relationships are solidified by a common purpose and a common enemy. In Bosnia, as in construction, competing interests jostled for position. Thinking back on Ludgate, I started to understand how competitive tendering affected human behaviour.

This insight was crystallised for me by a particularly memorable incident in Bosnia. I was driving through a forested area near Zenica with some colleagues when a man emerged from the woods waving his arms at the UN Land Rover. We were unarmed, so we had to be cautious. We signalled for him to stop a long way from the vehicle whilst we worked out whose side he was on. He turned out to be from Scotland. He was an electrician who had been working to build refugee accommodation for the UN. The project had come under fire, and the workers had only had one vehicle at their disposal. Not everyone, therefore, could be driven away from the fighting. They had agreed, in time-honoured fashion, that the married men would take the transport, and the rest had fled into the woods. As he explained the situation, his colleagues emerged from the trees. I knew three of them from Ludgate. They had all been left in a very vulnerable position, running from fierce crossfire, and they were only trying to do their job. I did what any decent human being would have done and paid for their flights home. In any case, I knew who they were working for, and imagined that, having got them out of a conflict zone alive, their employer would be only too pleased to square things up when I got back to the UK.

On my return to Britain, I went to see this highly respected trade contractor, hoping to get my money back. He looked at me as if I was mad.

'They're not our problem, mate,' he explained. 'They're self-employed subs.'

That was a low point in my perception of the construction industry. For me, it exemplified everything that was wrong with the attitudes created by the complexities of its contractual relationships. It was also the pivotal moment when I realised that to do things the way I wanted to do them I'd probably have to become a contractor myself.

I'd also learnt a lot about organisational politics in Bosnia – watching senior people trying to broker peace deals involved a similar power play to the negotiation of contracts – and thought I might be ready to start my own business, although I was less than enthusiastic about all the paperwork that this would entail. As it was, fate decided it for me. Just after landing at Heathrow on my return, I bumped into Ian Eggers, a Bovis manager I knew from Ludgate. Bovis was working at Heathrow for BAA. They knew they needed a construction logistics contractor, but such a company still didn't exist in the form we know today. Did I want the job?

Whilst I was thinking about this over the Christmas of 1993, I bumped into Mark Dobson, a lifelong family friend. I explained my dilemma. It turned out that he wanted to run his own business as well – and that he knew a lot about company administration. We agreed, over a few beers, to form a company. Feeling that 'Sullivan and Dobson' lacked a certain *je ne sais quoi* as a company name, we settled on Wilson James by sticking a pin in the phonebook. Many of the old Ludgate logistics team agreed to return, and most of those that did have remained with Wilson James to this day.

In the early days, we were lucky enough to get contracts from Bovis, Mace, BAA and Stanhope in reasonably quick succession. By this point, the company was established, and has grown from there. My professional involvement with my co-author, Stephen Barthorpe, also started to grow around this time. I had first met Stephen at Ludgate, after he wrote to Stanhope to ask if he could bring some construction students for a tour of the site. Peter Rogers had said yes, and the task of organising it had eventually been passed down the line to me. I arranged a couple of presentations for them, and delivered one myself on our experiments in construction logistics.

Stephen grasped the point immediately, and soon, in his capacity as a principal lecturer at the University of Glamorgan, incorporated logistics into an assessed module within their construction degree courses. I believe that was the first time anywhere that such an element had been formally included within a construction degree. One of his students, Stephen Robbins, went on to devote his MPhil thesis to construction logistics. I was delighted when he invited me to deliver some lectures to the students myself, and found helping them with their studies incredibly rewarding.

When I was at school, my teachers and I had never learnt to fully appreciate each other, and so I spent my early life unencumbered by any qualifications, aside from a swimming certificate. During my army days, however, I was selected to train as an instructor. I vividly remember the title of the first lecture: 'The promotion and maintenance of the desire to learn'. This was a new and exciting concept indeed. It was like a light-bulb turning on. I've been very pro-learning ever since. When I realised how many different qualifications my staff were obtaining, I went to the CITB, who gathered together other stakeholders and employers, and we created an NVQ Level 2 in construction logistics. For some members of staff that can represent their first ever formal educational attainment, and I believe that the commitment Wilson James has made to training has had enormous benefits for the whole company.

Still, I'm very conscious of the need to make learning as painless as possible. Therefore, I have deliberately kept my contributions to this book as straightforward and easy to engage with as I possibly can. Any claims to erudition that this book can make have been entrusted to the capable hands of my learned co-authors. However, I think I speak for all three of us when I say that I hope, above all, that it will simply be a book that you find useful.

Gary Sullivan

Section 1
Contextualising Logistics for Construction

Chapter 1
The Origins of Logistics

This chapter provides definitions and an overview of the origins and development of the concept of logistics. From its original military associations to the more universally applied supply chain resource applications, logistics has developed and become accepted as an integral component of modern business practice, particularly in the retail and manufacturing industries.

Definitions and origins of logistics

According to the *Oxford English Dictionary* (*OED*), logistics comprises the 'organization of supplies, stores, quarters, etc., necessary for the support of troop movements, expeditions, etc.' The *OED*'s first recorded usage of the term is 1879.

Further research, however, reveals that the term was in use at least seven decades earlier, at the time of the Napoleonic Wars. The year 1811 saw the publication of the informatively entitled *Elements of the Science of War Containing the Modern Established and Approved Principles of the Theory and Practice of the Military Sciences: viz the Formation and Organization of an Army and their Arms &c &c. Artillery; Engineering; Fortification; Tactics;* ***Logistics****; Grand Tactics; Castrametation; Military Topography; Strategy; Dialectic and Politics of War*, written by William Müller, an engineer.

A contemporary reviewer stated that 'in the composition of this work it has been Mr Müller's object to give a scientific view of the whole business of war from the mere handling of a spade up to the conduct of an army' (Anon 1812). The era which saw the mobilisation of unprecedentedly large armies across Europe had also produced widespread interest in all elements of the prosecution of wars, including the mechanics of supply, and classic military writers such as Jomini and Clausewitz were driven to explore new theories so that they could try to understand the battles that raged across the continent.

It is sometimes imagined that the word must have been derived from the Greek *logos*, meaning 'word', from which we get 'logistician', meaning 'someone who is skilled in logic'. Whilst an aptitude for logical thinking is undoubtedly a major asset to a logistician concerned with organisation and supply, the word, when used in this sense, actually comes from a completely different source. In low Latin, a *laubia* was an arbour, or temporary shelter made from trees. This became the French *loger*, and from thence brought words such as 'lodge' and 'logistics' into the English language (Chambers 1875). Whereas *logos*, the term for 'word' and 'reason', has always had a spiritual dimension in English – 'In the beginning was the Word', the Gospel of John begins – *loger* was always firmly rooted in the temporal. From its earliest forms, it has been a word to describe practical responses to changing contingencies. The exacting nature of the task of organising such responses is now so well established that the *OED* has recently drafted a new entry for 'logistical nightmare', citing Lord Mountbatten's description of Assam in 1944 as an early usage.

Whilst the word 'logistics' was long used to describe military situations, its usage in the world of retail and manufacturing is now well-established, although, as noted in the Preface, too often conflated with transport. Taylor (1997) advises that the comprehensive definition of logistics developed by the US Council of Logistics Management in 1986 is:

> *The process of planning, implementing and controlling the efficient, cost-effective flow and storage of raw materials, in-process inventory, finished goods and related information from point of origin to point of final consumption for the purpose of conforming to customer requirements.*

'Logistics' is defined by the Chartered Institute of Logistics and Transport (2006) in the UK as, 'the process of designing, managing and improving such supply chains, which might include purchasing, manufacturing, storage and, of course, transport'.

A word with its origins in temporary shelters made from trees has come to describe one of the most complex and demanding functions in modern commercial activity. In the following two chapters, we will trace that development, beginning here with its origins in the world of warfare. The military origins and application of logistics provide a fascinating insight which helps our understanding of how modern logistics has developed today.

The military origins of logistics

> *A real knowledge of supply and movement factors must be the basis of every leader's plan; only then can he know how and when to take risks with those factors, and battles are only won by taking risks.*
>
> Field Marshall Wavell (van Creveld 2004)

As the military historian John Keegan has observed, there are some basic constants to warfare. In Roman times, it was advised that soldiers should practise carrying loads of up to sixty pounds, whilst on the Somme, in 1916, sixty-six pounds was the average burden of a foot soldier (Keegan 1993). Little had changed over two millennia. He goes on to point out that there have been exceptions to this limit, such as the massive loads carried by the first British troops to land in the Falklands in 1982, but that even the fittest elite forces find it physically exhausting to carry such weight for any length of time.

Given the physical limitations of the soldiers themselves, the matter of how to provide enough food for soldiers fighting away from their own territory has been a problem ever since war ceased to be a fleeting and localised affair. As warfare became more sophisticated, and living off the land became problematic, supplies of foodstuff and other essentials had to keep pace with the advancing armies. Then there was the problem of maintaining essential supplies for both armies and livestock. These problems could partially be solved by placing additional loads on pack animals and wagons. These laden wagons became known collectively as the 'baggage train' and were normally placed at the rear of the marching column.

Until relatively recently, there were usually a large number of non-combatants who accompanied an army, especially during sustained and persistent invasions and campaigns. Even during raids, a substantial number of non-combatants, including carpenters, blacksmiths, tailors, armourers, fletchers, cooks, bakers and whores, accompanied the fighting forces. The non-combatants in a persisting force might have been more than double the number of combatants, therefore seriously exacerbating the supply problems.

A well-stocked baggage train allowed an army to be relatively self-contained as it carried its entire supply of requirements for the conduct of battle and the sustenance of troops and animals. However, the necessity of having a baggage train introduced its own distinct problems. The distance and speed at which an army could travel depended greatly upon the ability of the baggage train to deliver the supplies at the right place and time. This situation was not entirely satisfactory, as increases in the size of an army dictated a corresponding increase in the size of the baggage train. A large, ponderous baggage train was also susceptible to enemy attack.

Two methods of acquiring supplies on the move were therefore developed for use when a baggage train was unsuitable or impractical. The first was to purchase (or take) supplies from people living near to or along the army's route of march. The alternative was to stockpile supplies at fixed, fortified bases along the route of march and either bring them forward by wagon to the army as required or collect them as the army marched by.

The Romans were considered the first acknowledged great 'global' power to use a combination of all three systems quite successfully. The

Roman legion's ability to march fast was attributed to various factors, such as superb roads and efficiently organised supplies. Their supply trains featured mobile repair shops and a service corps of engineers and armourers. Supplies were usually obtained from local sources and stored in fortified depots.

For instance, the Scottish campaigns of the Roman emperor Septimius Severus at the beginning of the third century saw the Roman fort at South Shields, at the mouth of the river Tyne, being converted into a supply base (Johnson 1983). The original interior was demolished, and all that was retained of the fort's original structures were the headquarters' building and the fort's double granary. The rest of the interior was replaced by no fewer than 18 single granaries. The garrison itself was moved into an annexe. Grain would have been imported by sea, perhaps from as far away as Holland. Long before the advent of the modern world, large-scale international solutions could be engineered when it suited the military aims of a major power.

However, it should also be remembered that most early armies marched on foot, and that the poor quality of the roads, which were often little more than tracks, meant that travelling by land was enormously difficult. The exception was the great routes carved out by the Roman Empire. When that empire, and its roads, collapsed, 'that decay spelt an end to strategic marching everywhere for more than a thousand years' (Keegan 1993).

The leaders who realised the importance of finding solutions to these limitations had a strategic advantage. Alexander the Great benefited from the innovations of his father, Philip II, who had realised that ox carts were undesirable because of their slow speed and poor endurance over long distances. Instead, he reduced the train by increasing the loads of his soldiers as much as possible and banning women and children from following the army. This 'made the Macedonian army the fastest, lightest, and most mobile force in existence, capable of making lightning strikes ... Alexander's astonishing speed, which so terrified his opponents, was due in no small part to Philip's reforms' (Engels 1980).

Still, the army could not stay self-sufficient for long, unless near a sea port or navigable river. Therefore, Alexander always garrisoned his troops in well-stocked, populous areas until the winter harvest was gathered, and tried to avoid moving his entire army into an area before a surrender – and therefore supplies – had been successfully negotiated (Engels 1980). Plunder has always featured heavily in warfare, but few areas, willingly or otherwise, can support a sizable army for long, particularly in winter.

Another later example of logistical skill as a military advantage is that demonstrated by the Mongols, who, in the thirteenth century, derived much of their strength from developing excellent supply systems. Their cavalry armies had one of the most efficient logistical systems known. It was based on self-containment and local supply supplemented by bases established at strategic positions. In normal movements, the Mongol armies divided into several corps and spread over the country, accompanied by trains of baggage carts, pack animals and herds of cattle. Routes and campsites were carefully selected for accessibility to good grazing and food crops. Foodstuff and forage were stored in advance along the routes of march. The Mongols conquered extensively from Europe to the Pacific, demonstrating the value of their efficient logistics system.

During the medieval period, however, there was little development in the theory of logistics. Commanders tended to rely on the thinking of the ancients, continuing a tradition whereby 'war consisted mainly of an extended walking tour combined with large-scale robbery' (van Creveld 2000). The history of logistics is one of negligible progress followed by rapidly accelerating change in the wake of technology and new thinking.

Between the wars of the mid-sixteenth century and the Battle of Waterloo in 1815, significant changes occurred in strategy, tactics and technology. This military revolution was initiated by the Spanish in the Spanish Netherlands during the Eighty Years War of 1568 to 1648. During this conflict, the infantry gained ascendancy over the cavalry and standing, professional armies were founded. The size of the battlefield expanded, armies increased in size and greater control over the army was exercised by centralised bureaucracies and monarchies.

While the supply problem had remained essentially the same as in the fourteenth century, revised supply and transportation methods had to change in order to cope with changes in military tactics. The Spanish were deprived of sea supply routes by the Dutch and had to develop sophisticated overland supply routes from Spain to the Netherlands using supply stations en route a day's march apart. A contractor at each station, hired by Spanish officials, was responsible for providing sufficient supplies. Spanish garrisons in Flanders had their food supplies and other necessities provided by contractors and private entrepreneurs in a concept known today as 'contracting-out'.

The magazine concept (known today as a 'primary distribution centre' or 'consolidation centre') used by the Spanish evolved during the reign of Louis XIV of France (1643–1715). There were two underlying principles to the magazine system. First, a ration scale was developed to indicate exactly what each member of the army was entitled to. Second, there were standard contracts for the provision of the required services, whereby contractors supplied direct to the army or at an agreed established supply base.

Also during this period, a number of books came out on the theory of war, including a notable contribution by an Italian called Raimondo Montecuccoli. He helped to define what society came to understand as war – primarily a conflict between states rather than an internal uprising or insurgency. However, he 'still failed to distinguish between strategy, the operational level and tactics' (van Creveld 2000). This meant that he and his eighteenth-century successors tended to work from the 'bottom up' (van Creveld 2000), focusing on technical operations and building tactics around them, rather than using logistics to support a strategic goal.

During this period, gunpowder was increasingly used, and despite the difficulty of supplying its vital constituent of saltpetre (there being no synthetic way of producing it until the mid-nineteenth century) infantry armed with pikes had all but disappeared in Western warfare by the end of the seventeenth century. As the method of waging war changed forever, the rise of the musket and cannon meant that the supply of ammunition powder and muskets became paramount for the conduct of modern warfare. More than ever, supply trains became an absolute necessity. This change also introduced a new element of distance to warfare – aside from a contingent of archers, war had previously been a hand-to-hand melee. It also meant that infantry could be trained far quicker. Mastering the longbow was a lifetime's study for the medieval archer, whereas a musketeer could be trained in a matter of weeks.

Just as gunpowder introduced an element of distance to battles, the development of cartography to the point where, by the end of the eighteenth century, physically accurate maps became available meant that operations could sensibly be planned at a distance as well. This would have an impact on the development of military thought, notably the contribution of Adam Heinrich Dietrich von Bülow (1757–1807).

Bülow understood the rising importance that firearms had accorded to logistics, and theorised that military advantages could be obtained by spreading out supplies so that they were being brought to the front via multiple lines of communication. He dealt with this in a highly mathematical way, arguing for converging lines that were spread out geometrically from a wide base. This was intended to make the disruption of supply lines by the enemy as difficult as possible (Gat 2001).

Indeed, Bülow believed that such strategies could ultimately render actual battles totally unnecessary, since a retreat could be forced merely by cutting the supply line. 'War will be no longer called an art, but a science,' he said. He believed that through the application of a set of rules 'the sphere of military genius will at last be so narrowed, that a man of talents will no longer be willing to devote himself to this ungrateful trade' (Bülow quoted in Gat 2001).

Unfortunately, Bülow's logical and abstract approach to the prosecution of war took little account of geographical peculiarities or other situational factors. Paradoxically, this theory of warfare was both rooted in logistics and incapable of properly accounting for it. This failing became painfully apparent through the example of his contemporary, Napoleon Bonaparte, who was bringing a new vigour to war across Europe, without apparently resorting to Bülow's geometrical approach to supply.

Napoleon knew that an army marches on its stomach – he was the originator of the expression. In his early campaigns, he had been forced to rely on local levies and loot from captured areas. He was also acutely aware of the challenges good military logistics present. '*Qu'on ne me parle pas des vivres*' (Let no one speak to me of provisions), he is reported to have said. However, the French had already developed a culture of approaching logistics in a systematic manner: Louis XV had created a whole department under a *maréchal des logis*. This marshal of logistics role translates broadly into what we would know as a quartermaster, with responsibility for lodgings and quarters. However, after the French Revolution, the old term of *logis* was superseded by *logistique*, reflecting a new sense of the importance of a scientific approach to the challenges of military supply.

Napoleon also moved the focus of campaigns from besiegement to the decisive engagement with the enemy's field army. This continued a process, begun with Montecuccoli, of developing what has been termed 'trinitarian warfare' (van Creveld 2000). State engaged with state, army with army. The civilian population was now cast in a largely passive role. The great theoretician Karl von Clausewitz saw war as the 'continuation of politics by other means' (Keegan 1993), and so it must have seemed in an era of the rising nation state and the development of orderly rules of engagement.

As the nineteenth century wore on, there was also acceleration in the development of technologies that would be used for war-making. Tinned food made its first appearance, and in 1855 the first railway to be built for a military purpose was laid at breakneck speed in the Crimea by Sir Samuel Morton Peto. As the navvies arrived to build it, the future General Gordon, then a young lieutenant, wrote, 'No relief that could be named will be equal to the relief afforded by a railway' (Peto 1893). Indeed, it has been stated that it 'could be convincingly argued that Peto and his navvies did far more to alleviate the suffering of British troops in the Crimea than Florence Nightingale and her ladies' (Collier 2007).

The railway would henceforth only increase in importance. The Prussians won military victories against Austria (1866) and France (1870) through rail superiority, and possession of a railway network gave the North a crucial advantage in the American Civil War. By 1914, millions of troops were being mobilised in a matter of weeks through the railway system. The ammunition being transported by new mechanised means (the motor vehicle, as well as the railway, would now be part of future war efforts) also rose exponentially. Napoleon had 246 guns at Waterloo. Only a century later, a million artillery shells were fired by the British in the week leading up to the Somme (Keegan 1993).

The First World War, however, was both the epitome of logistical achievements at that date and a grave sermon on the limits of logistical operations as strategy for achieving military objectives. Although new technologies such as the tank and the aeroplane were being developed, they had not yet reached a level where they could make a decisive contribution. According to Keegan (1993), 'The First World War was eventually resolved not by any discovery or application of new military technique by the high commands, but by the relentless attrition of manpower by industrial output ... ever wider recruitment of men, ever costlier purchases of arms had cancelled each other out. Supply and logistics had damaged all the combatants in almost equal measure.' Conversely, World War II would be settled decisively through 'industrial output', as American production outpaced all efforts from her enemies to sabotage them in transit.

Modern military logistics

Hamilton (2003) states an old military adage, 'Amateurs talk strategy; professionals talk logistics' and argues that, once the strategists determine the aim of the war and the generals set the size and shape of the force needed to accomplish it, the logistics commanders must get all the pieces to the war theatre. Wars are planned and fought with a combination of logistics and tactics.

Centralised organisation evolved from the military system of logistics to provide overall guidance, a civilian management team that contracted for the supplies, a disciplined transport system that delivered supplies to the troops and a group of specialist staff officers who were concerned with making the system work – the specialist logistics manager was born. In 1993 the British Army formed its largest corps – the Royal Logistics Corps (colloquially known in the service as the 'Really Large Corps' or 'The Loggies') – by the union of the:

- Royal Corps of Transport
- Royal Army Ordnance Corps
- Royal Pioneer Corps
- Army Catering Corps
- Royal Engineers Postal and Courier Service.

The Royal Logistics Corps is responsible for the provision and distribution of equipment and stores for the Army, from mail to ammunition, by rail, road, sea and air. It also provides catering and specialist bomb disposal services to the Army (British Army 2006).

In addition to the practical advantages to the British Army for implementing dedicated and modern logistics support there is increasing commercial pressure from the UK government Treasury to make military logistics more cost-effective and efficient. Defence-related inflation has for many years exceeded normal economic inflation, according to *The Economist* (1998), and implementing 'lean supply chain' principles into 'lean logistics' to reduce costs is seen as an imperative. Although modern warfare can benefit from utilising modern commercial principles of 'lean supply chains', 'value streams' and 'just-in-time' techniques, the dilemma is in balancing the principles of 'just in time', 'just enough' and 'just in case' with the overwhelming imperative to win the war.

Antill (2001) suggests that with the huge technological advances of the twenty-first century and the need for armies to engage rapidly far from home, the military is adopting some of the best practices of the business and commercial worlds. The United States Armed Forces have evolved 'lean logistics' principles into a concept known as 'focused logistics', which their Department of Defense defines as, 'Logistics of the future, just as in the past, will continue to play a crucial role in our nation's force projection capability. The ability of the nation to project and sustain power in the future will require a fusion of information, logistics, and transportation technologies to provide rapid crisis response, track and shift assets even while en route, and deliver tailored logistics packages and sustainment directly at the strategic, operational, and tactical levels of operation' (United States Transportation Command 2009). The emphasis of focused logistics is to provide 'total asset visibility to enable logistics to be based on velocity of distribution rather than stockholding'.

Military logistics has developed significantly from lessons learnt during the Falklands War and the two Gulf wars. The combination of factors such as the distances from home bases, harsh climates and the huge amount of supplies needed demanded a level of efficiency unequalled in previous conflicts. According to Arabe (2003), lessons learnt following the first Gulf war and implemented by the US military in the second Gulf war included improvements in 'asset visibility' featuring radio-frequency identification (RFID) technology to keep track of supplies and provide accurate information to the troops regarding expected arrival dates. Improvements to inventory control and procedures also ensured that for the second conflict detailed information about the supply containers was known prior to their shipment, obviating the need to open the containers upon arrival to determine their contents.

Parallel lines: construction and the military

In 1815 Napoleon was defeated in the decisive battle of Waterloo. In the same year, the London builder Thomas Cubitt signed a contract to build premises for the London Institution (Hobhouse 1995). The contract was for £20,000, a massive amount of money, but the penalty clauses for late delivery were stiff. Thomas decided that he couldn't risk being let down by subcontractors, and hired his own men for the job. That is traditionally held as being the beginning of major contracting, a phenomenon which would only grow as the nineteenth century progressed.

In a sense, both warfare and construction had become more straightforward. Big armies and big contractors brought the strength and delineation of a major organisation to their respective endeavours. Large institutions dominated the pursuit of both creation and destruction, and would do so for almost 150 years. Contractor and commander alike held sway over resources and manpower, and it was easy to identify allegiances.

By 1945, however, both fields had started to change profoundly. The Second World War had wreaked medieval barbarism on civilian populations, and the War in the Pacific had been decided with the atomic bomb, a weapon so terrible that war-making would change forever. The major industrial powers could no longer countenance a major confrontation with their conventional armies, largely resigning war to minor powers and the auspices of the United Nations. As Keegan points out (1993), the triumph of the Western way of warfare was delusive – European tactics had subjugated almost the entire globe – but, 'Turned on itself it brought disaster and threatened catastrophe [and] terminated European dominance of the world.' Thereafter, war would become a steadily more fractured and ambiguous activity.

At the same time, the UK construction industry underwent a bloodless revolution, which would also change future tactics and reintroduce older customs into a world of ever-increasing technological complexity. As will be discussed in Chapter 4, the piecework agreement of 1947 ushered in an unforeseen growth in self-employment and subcontracting, bringing disjointed working relations to an arena which had once been far more clear-cut.

Both the military and the construction industry, therefore, were faced with a need to rethink their provision of logistics within a changing environment. The British government were acutely aware that logistical failings could easily have lost them the Falklands War – a conventional war fought in a remote and unconventional environment – and by the late 1980s they were considering the changes which would usher in the Really Large Corps. At the same time, Peter Rogers, Construction Director of Stanhope, realised the new form of contract, Construction Management, required a rethink of resourcing and supplies. Ten years later, Egan was calling for lean construction, and defence logistics were creating lean supply for the army.

Construction is often compared to the car industry, but it is argued here that it actually has far more in common with the military. Both domains are increasingly technology-driven, and yet war and construction still both depend on the bravery and determination of their foot soldiers for success. There have been years when more lives are lost on British construction sites than within the British Army, and both soldiers and construction workers rely on camaraderie and black humour to get them through a campaign (or project).

Whilst both these traditionally masculine callings remain taxing and sometimes dangerous, they are slowly broadening their constituency and improving conditions for their workers. There are soldiers in Iraq and

Afghanistan whose lives have been saved by modern body armour, just as construction workers are protected with fall-arrest systems. There are still casualties, but the march of progress continues to reduce unnecessary suffering.

The importance of getting the right equipment to the place it's needed, whether it's the vehicles the army needs to withstand improvised explosive devices or construction's manual-handling equipment, can therefore be, quite literally, a matter of life and death. The military has embraced the challenge, and now construction is starting to do the same. Change was always going to be quicker in the military – construction has no point of central command – but the industry is now waking up to the need for focused and practical solutions to its logistical challenges. The smooth logistical operations of the car industry can probably never be replicated in the muddy and messy world of construction, but, like the military, it can give its troops the best possible environment and equipment to perform a demanding but vital function for society.

References

Anon (1812) Article VIII. *The Eclectic Review* Volume I.

Antill P (2001) Focused logistics, http://www.historyofwar.org/articles/concepts_focused_logisitics.html, [accessed 7th January 2010].

Arabe KC (2003) Logistics lessons from the military, http://news.thomasnet.com/IMT/archives/2003/06/logistics_lesso.html, [accessed 3rd January 2010].

British Army (2006) Key facts about defence, http://www.mod.uk/Defenceinternet/About/Defence/Organisation/KeyfactsAboutDefence/TheBritishArmy.htm, [accessed March 2006].

Chambers W (1875) *Chambers's Etymological Dictionary of the English Language*, Chambers, London (reprinted by Elibron Classics, 2005).

Chartered Institute of Logistics and Transport in the UK (2006) http://www.ciltuk.org.uk, [accessed 14th February 2006].

Collier C (2007) *CIOB Construction Leaders*. Chartered Institute of Building, Ascot, www.ciob.org.uk/filegrab/BOOKLETv15FINALwebspreads.pdf?ref=408, [accessed 6th January 2010].

van Creveld M (2000) *The Art of War: War and military thought*. Cassell, London.

van Creveld M (2004) *Supplying War: Logistics from Wallenstein to Patton*, 2nd edn. Cambridge University Press.

The Economist (1998) 'Platform envy', 12th December 1998, p. 25.

Engels D (1980) *Alexander the Great and the Logistics of the Macedonian Army*. University of California Press, Berkeley.

Gat A (2001) *A History of Military Thought: From the Enlightenment to the Cold War*. Oxford University Press, Oxford.

Hamilton D (2003) Logistics dictate Iraq war timing, http://www.globalsecurity.org/org/news/2003/030204-iraq01.htm, [accessed 3rd March 2006].

Hobhouse H (1995) *Thomas Cubitt: Master Builder*. Mercury Business Books, San Francisco.

Johnson A. (1983) *Roman Forts of the 1st and 2nd Centuries AD in Britain and the German Provinces*. Adam & Charles Black, London.

Keegan J (1993) *A History of Warfare*. Pimlico, London.

Müller W (1811) *Elements of the Science of War Containing the Modern Established and Approved Principles of the Theory and Practice of the Military Sciences: viz the Formation and Organization of an Army and their Arms &c &c. Artillery; Engineering; Fortification; Tactics; Logistics; Grand Tactics; Castrametation; Military Topography; Strategy; Dialectic and Politics of War*. Longmans, London.

Peto H (1893) *Sir Morton Peto: A memorial sketch*. Elliot Stock, London.

Taylor D (1997) The analysis of logistics and supply chain management cases. In: *Global Cases in Logistics and Supply Chain Management*. International Thomson Business Press, London.

United States Transportation Command (2009) USTRANSCOM Fact file: Future logistics requirements, http://www.transcom.mil/factsheet.cfm, [accessed 11th January 2010].

Section 2
Construction Logistics in Practice

Chapter 2
Construction Logistics in Practice

This chapter describes the origins and development of logistics as a specific discipline in the construction industry. The traditional approach to logistics is compared to the advent of the dedicated approach, and this chapter provides compelling reasons for the adoption of the latter approach.

Development of logistics

Logistics has developed into a relatively sophisticated and integrated management process. Taylor (1997) suggests that since the mid-1970s there has been a significant development in the theory and practice of logistics management and a greater degree of understanding of how supply chains operate. Taylor contends that 'logistics management' and 'supply chain management' (SCM) are essentially synonymous terms involving the systematic and holistic approach to managing the flow of materials and information from its raw material state to the end-user's consumption. Concomitant with the philosophy of SCM has evolved the concept of just-in-time (JIT) deliveries. Traditionally, many organisations would hold considerable stockpiles of components prior to their eventual use in the manufacturing process. This has proved to be an expensive and inefficient use of finance and space resources and has become an obsolete practice in favour of the JIT concept pioneered by companies like Toyota in Japan.

The Chartered Institute of Logistics and Transport in the UK (CILT UK) was established in 2004, although it has its roots in the Chartered Institute of Transport, which was formed in 1919 and was granted its Royal Charter in 1926. The original Royal Charter of the Institute of Transport quotes the objectives of the Institute as including:

> *To promote, encourage and co-ordinate the study of the science and art of transport in all its branches.*

CILT UK's mission statement is:

> *To be the focus for professional excellence, the development of the most relevant and effective techniques in logistics and transport, and the development of policies which respond to the challenges of a changing world.*

Effective logistics management is an integral and critical success factor in differentiating competing organisations and even competing supply chains in this increasingly global, competitive economy. Logistics, according to CILT UK (2006), sets out to deliver exactly what the customer wants – at the right time, in the right place and at the right price. Very often transport is a major component of the supply chain which delivers to the customer the goods and services needed.

The advent of warehouse management systems (WMS) and a range of Internet-enabled computerised document control and satellite navigation vehicle tracking systems have provided supply chain partners and their customers with sophisticated management support systems to increase efficiencies and certainties and decrease costs and time.

Construction industry logistics

Although there are numerous examples of best practice and even world-class performance, the construction industry, apart from those aspects associated with major commercial and infrastructure projects, has generally been slow to recognise the advantages of using dedicated logistics professionals and strategies. The acceptance and implementation of specialised and integrated logistics into mainstream construction is yet to be widely adopted in the UK despite its proven track record of cost reduction and resource efficiencies in other industries.

Logistics as a construction discipline first came to prominence in the UK in the 1980s with the advent of construction management, a procurement mechanism whereby the client directly contracts with specialist trade contractors, albeit still using a project management contractor to coordinate and manage the construction process. Some of the clients that utilised the construction management option recognised the value of directly engaging specialist firms to manage the essential non-core construction activities, usually cited as 'preliminary items' in contractual documents, including site welfare, housekeeping, safety signage, security, fire, access, cranes, hoists, traffic management, first aid, third-party liaison and waste management. The amalgamation of these essential, albeit non-core, elements into one package led to the emergence of the specialist logistics contractor.

Seminal reports and reviews on the UK construction industry (Latham 1994; Egan 1998; Bourn 2001; Strategic Forum for Construction 2002) have lamented the inefficiencies and waste due to poor logistical performance. The *Improving Construction Logistics* report published by the Strategic Forum for Construction Logistics Group (2005) argues that there is a lot of opportunity for change and that the construction industry has been slower than other industries to realise the benefits that the application of good logistics can provide.

This book examines the failures of the existing practice adopted by the construction industry and considers the merits of alternative, modern and integrated approaches to logistics.

British Airports Authority (BAA) is one of the UK's biggest construction clients and former BAA Group Technical Director, Tony Douglas, contends that 'logistics is going to become absolutely crucial to construction. Other industries have improved working capital by 20% by getting smart on logistics' (Fairs 2002).

Well-established logistics specialist companies in the UK, like Wilson James, have already made a significant impact on improving the construction industry by introducing effective logistics management processes to otherwise traditional industry practices. According to Fairs (2002), logistics professionals and global logistics firms like Exel and UPS are also targeting the construction industry, as they claim that contractors could reduce their materials and labour costs by up to 15% by introducing JIT deliveries to sites. This would also reduce damage to materials, save on storage costs and free up skilled workers for more productive work.

An increasingly competitive marketplace has focused industry leaders' attention on maximising site-based productivity and minimising wasteful activity in the entire construction process, including the delivery of materials and equipment to site. The Strategic Forum's report *Accelerating Change*, published in 2002, argues that there 'should be greater focus on SCM and logistics to facilitate the elimination of waste'. The method and efficacy of how construction projects are delivered is also attracting more attention, with legislation such as the Town and Country Planning Act (1990) and initiatives such as the Considerate Constructors Scheme (2006) increasing contractors' accountability, in an attempt to improve the image of the industry. In addition, construction clients are becoming increasingly aware of the importance of their corporate image and are demanding that more consideration be given by their contractors to the needs of society.

The manufacturing and retail industries have developed their SCM and logistics techniques to good effect and benefit from improved efficiency and profitability as a result. When compared with these, the construction industry's ad hoc approach is far more simplistic and leaves considerable scope for improvement. Construction logistics and SCM are areas that have a significant effect on both project and industry performance, and yet have remained largely unchanged over many years. An estimated 10 to 20% of all construction costs are transport-related, according to the Building Research Establishment (2003), and approximately one in every five vehicles on UK roads is on construction-related business. Despite this, little consideration is given to the method by which materials and equipment are delivered to site; as a result, waste is increased and cost certainty is reduced throughout the supply chain.

Logistics involves the strategic and cost-effective storage, handling, transportation and distribution of resources. It synchronises parties in the supply chain from the source of origin to the point of use, with the aim of meeting the needs of the end-user. It is a process capable of being managed and is normally aligned with key times/dates in order to support the primary business activity.

Comparing construction with manufacturing

The *Rethinking Construction* report (Egan 1998) recommends that manufacturing industries be investigated to inspire new and improved working practice in the construction industry. The UK manufacturing industry has identified the benefits of developing a reliable and robust logistics strategy and focused upon achieving this over the past 30–40 years.

Although there are many similarities between construction and manufacturing, it is impossible to entirely replicate the methodology, because of fundamental environmental, cultural and operational differences between the two.

The factors which determine the success of a logistics activity have previously been identified as being; time, location, quantity and quality. To consider the potential for the construction industry adopting the manufacturing industry's approach to production a comparison of the environments has been made based on these four elements. This includes the interpretations of deliveries to the point of use, JIT and the use of work packs.

Deliveries to the point of use

In manufacturing, 'point of use' literally means the location where the 'fitter' installs a component into a machine. Materials are taken from the 'stores' to the production line by the 'line feeder' and placed within an arm's length of the fitter. This process forms a small but essential element of a factory logistics strategy and eliminates the need for the fitter to leave the production line. In construction, this interpretation of 'point of use' is often not possible, because of operational or safety reasons which might include working in confined spaces or where the task preceding the delivery requires setting time, for example a floor screed.

The conveyor-belt, production-line system is the key difference between a manufacturing plant and a construction site as it provides a defined and static point of use by taking the product to the fitter. In addition, the manufacturing strategy complements the use of the conveyor system and allows a fitter to have a narrowly defined task, for example the installation of a single component. This increases the level of certainty when planning, as it is possible to predict the precise number of tasks a person is able to complete during a given period of time, their tool requirements and to structure other operations accordingly.

Just-in-time deliveries

The manufacturing industry's interpretation of JIT is very literal. Materials are 'picked' from the stores, taken to the 'line' and used within minutes. Again, this level of efficiency is complemented by the layout of the factory and the production process.

The planning process used in construction is affected by external influences such as weather conditions, environmental factors and site topography and therefore does not offer the same level of certainty as the planning process in a manufacturing environment. Consequently, the precise timing of deliveries is very difficult to achieve and is compounded by the variety of tasks a construction tradesperson performs throughout a day.

Work packs

Furniture retail has made particularly good use of flat-pack units. Manufacturers commonly despatch packs containing each item required to construct a particular piece of furniture, such as a wardrobe or kitchen suite.

This method of furniture retail makes it easier for the purchaser to construct the product in the home, guarantees that the components used are compatible and reduces the build time by eliminating the need for the purchaser to locate/procure components and/or tools to complete the task. However, providing a complete pack requires the designers to consider the construction process in great detail and involves greater upstream logistical participation as structural components, fixtures and fittings all need to be included off-site.

Although manufacturing industries can be used to inspire new working practice in terms of off-site fabrication and elements may be transferable, it is not possible to entirely replicate each of its elements on a construction site, because of the following reasons:

- The controlled factory environment usually benefits from having an ergonomically designed layout and the facility to provide employees with specifically defined roles that enable greater certainty when planning.

- The lifespan of a construction project is extremely short compared with the lifespan of a manufacturing plant, therefore investment in innovation is not considered justified.
- The location is also a major factor. The manufacturing industry has traditionally settled in areas where land prices and cost of labour are low, whereas construction projects that warrant the use of a consolidation centre are likely to be in areas where the level of demand on land and property is high.
- In manufacturing, the carefully considered use of ergonomics, use of a conveyor system, the production of a standard product and the limited range of individual tasks on a production line focus efforts on maximising productivity. Whereas construction work is relatively short-term and is often sited in particularly difficult circumstances. Most buildings are custom-designed, with each tradesperson performing a multitude of tasks and, because of its comparatively short lifespan, the consideration made to ergonomics is considerably less.
- Due to the extensive range of materials handled and the knowledge required to prepare 'work-packs' in construction, its application is limited without the involvement of the upper end of the supply chain and consideration during the design stage.

Figure 2.1 illustrates that logistics is an important element to any process involved in the production of a complex, multi-component product, such as the construction of a building or the manufacture of a car, as it synchronises parties in the supply chain. This is particularly true for time-critical operations, as greater emphasis is placed on the punctual availability of materials, resource and information. Therefore, the success of any logistics activity can be determined by its ability to make resources available to the next user in the supply chain and ultimately the end-user. To achieve this, time, location, quality and quantity are essential.

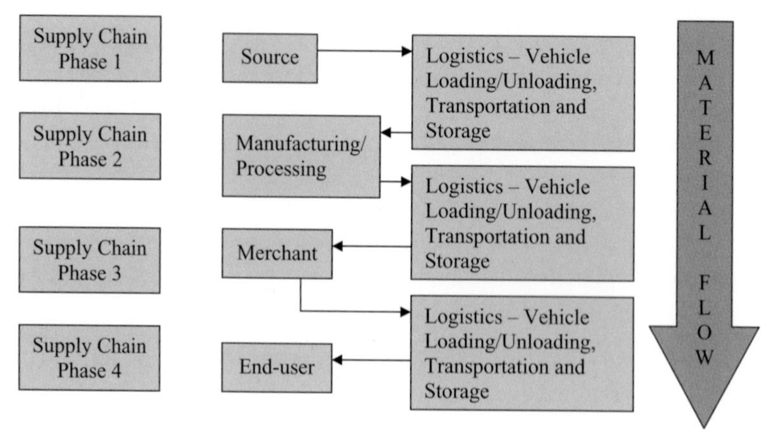

Figure 2.1 The interface between logistics and supply chain management

Construction industry: traditional approach to logistics

Logistics has the potential to affect shareholders and stakeholders alike. It is one of the most important elements of a construction project, influencing critical site performance factors such as cost, speed of construction and plan reliability, and industry performance indicators such as accident statistics and contributions to landfill. As a result of the significance of these issues, it might be expected that considerable attention be paid to logistics when developing a construction strategy. However, the reality is that little is known about construction-specific logistics and many of the issues identified in Figure 2.2 are not recognised or associated with logistics.

Traditionally, little attention has been paid to SCM or logistics and the industry has only recognised the final leg of materials delivery as being important. The plethora of 'forms of contract' available to procure construction work and a tendency to outsource risk have fragmented the supply chain. Each trade contractor is increasingly responsible for the procurement of their own materials. This has led to an unsatisfactory situation whereby no one has overall control of a project, which risks other stakeholders being adversely affected and project performance being diminished. Figure 2.3 illustrates the traditional, inefficient and environmentally damaging method of delivering materials to site.

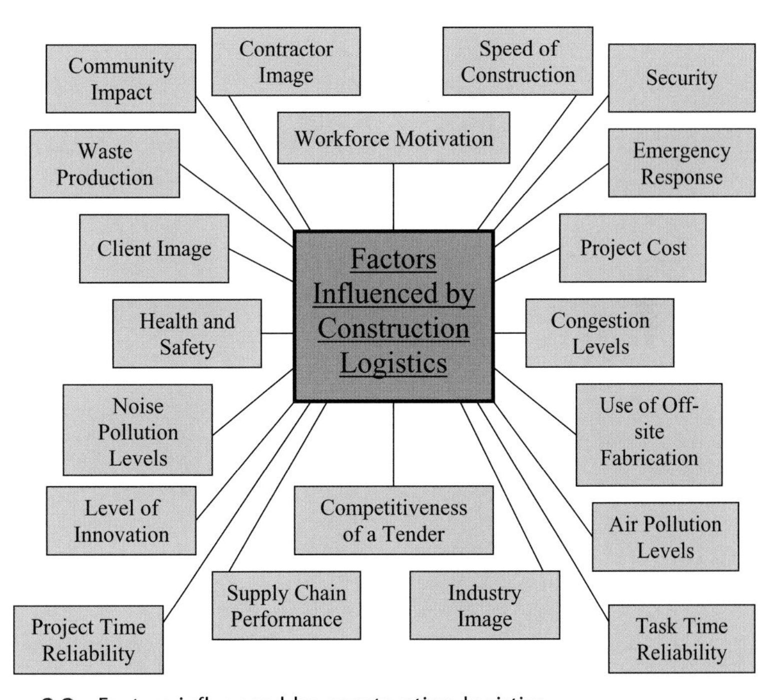

Figure 2.2 Factors influenced by construction logistics

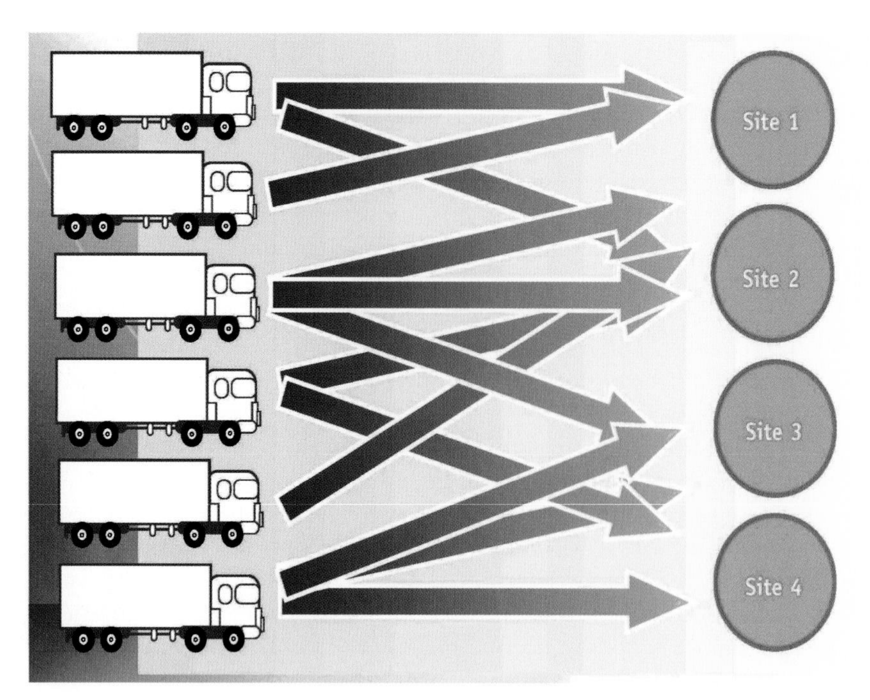

Figure 2.3 The traditional, inefficient and environmentally damaging method of delivering materials to site

Traffic congestion and limited storage space exacerbate this problem for sites located in city centres. In an attempt to manage the delivery process many sites stipulate delivery times; however, the methods employed by suppliers to meet these requirements involve the delivery vehicle arriving up to an hour early, necessitating the vehicle to either park outside the site or kill time by driving around the adjacent roads consequently adding to the problem of congestion and environmental pollution.

Upon arrival, materials and equipment are often offloaded and checked by the respective tradespeople, who are invariably over-paid, over-qualified, unmotivated, under-trained and ill-equipped to undertake this role. Alternatively, temporary labour, again lacking appropriate training, motivation or processes, is hired at short notice to unload materials. This is often done in a hectic site environment where there is fierce competition for space and resources. As a consequence, tradespeople are disrupted for a considerable period on each occasion. The situation is dangerous and hampers progress. In addition, a 'push based' delivery schedule along with inadequate storage facilities can result in materials being damaged. Figure 2.4 identifies many of the problems associated with the traditional approach to construction logistics.

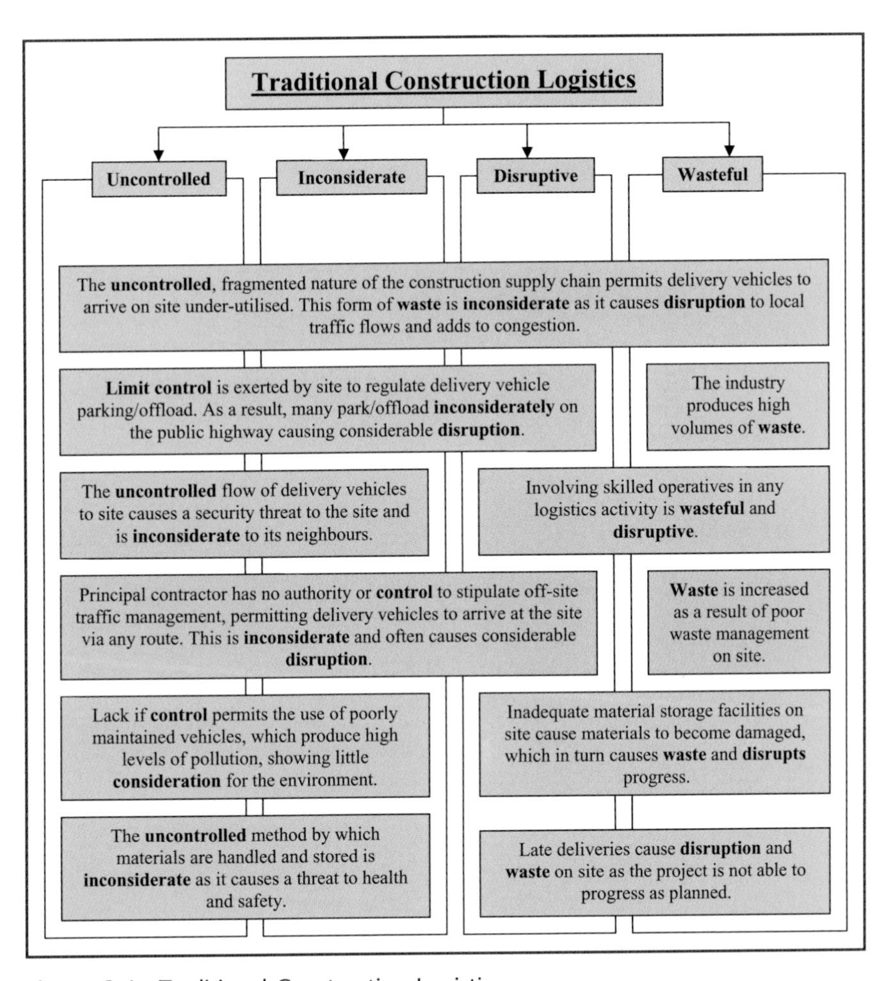

Figure 2.4 Traditional Construction Logistics

Construction industry: dedicated approach to logistics

Dedicated construction logistics has developed into an essential support service that adopts many of the characteristics, normally associated with facilities management. This has led to the emergence of the dedicated logistics contractor, who assumes single point responsibility to integrate all the essential support services associated with construction projects. Figure 2.5 demonstrates the typical role of a logistics contractor, identifying its primary and secondary responsibilities.

There are four compelling reasons why the construction industry should implement a dedicated approach to logistics. First, to maximise the productivity and efficiency of the skilled workforce on site (the same motives as in the manufacturing and retail industries). Second, to maximise quality of service by enabling a dedicated, trained logistics service team to provide a holistic support service for the construction project.

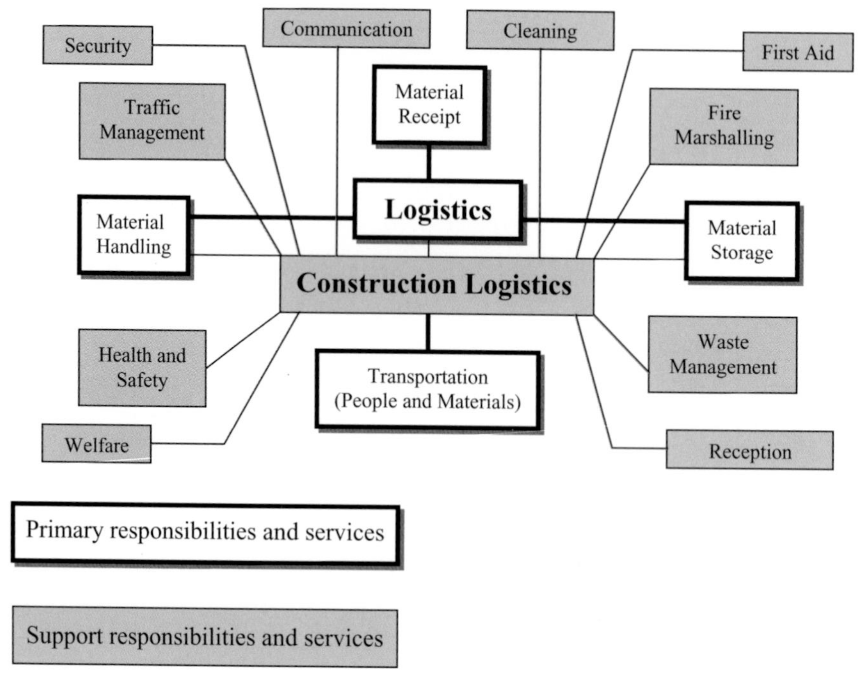

Figure 2.5 Primary and support logistical responsibilities and services

Third, to minimise the negative environmental and social impact that construction projects create and, fourth, to ensure that the highest possible standards of health and safety are attained by using an appropriately trained and experienced logistics workforce.

Maximising the productivity of the skilled construction workforce was the main driver for the primary development of construction-specific logistics. It was a logical waste-reduction exercise achieved by eliminating the need for skilled construction personnel to undertake non-construction work like materials handling and waste removal. For this, a dedicated team was made responsible for receiving, storing and distributing materials to the point of use. It was soon found that this method improved the standard of health and safety and reduced waste, as a result of more proficient handling techniques.

A secondary development (illustrated in Figure 2.5) demonstrates how a logistics team has extended their responsibilities on site to provide a more supportive role. This diversification means that the logistics contractor provides a multi-service team, capable of fulfilling a variety of roles. For example:

- As the logistics team was responsible for receiving deliveries, it was logical that it was also made responsible for supervising traffic whilst on site. As a result, many of the logistics contractor's employees became trained banksmen.

- Financially, it was more expedient for logistics operatives to be responsible for the general management of welfare facilities and site cleaning than it was to force subcontractors to do it or employ specific cleaning operatives to carry out such tasks.
- As logistics operatives were responsible for cleaning the site and had access to mechanical resources, such as forklift trucks, their duties were naturally extended to include materials handling and waste management.
- The logistics operatives are trained in first aid and fire marshalling duties, and so the site management team always had essential emergency/safety personnel present.

Although the site-based dedicated logistics team improves logistics on site, it has little effect on off-site logistics or supply chain performance, which is where many of the problems identified in Figure 2.4 (traditional construction logistics) exist. Improving off-site logistics was identified as a key issue in the seminal reports *Rethinking Construction* (Egan 1998) and *Accelerating Change* (Strategic Forum for Construction 2002). The reports also recommend that designers, contractors and product suppliers examine how to improve supply chain tracking and delivery systems through the supply chain to improve productivity. The construction industry should learn from the retail sector, where logistics has developed and become more sophisticated. It is a key area where the national retailers in the UK like Sainsbury's and Tesco compete against each other and where huge productivity gains have been achieved by overhauling their supply chains.

The scale of the environmental and social impact that construction projects create is significant. A mass balance study of resource use, wastes and emissions in the UK construction industry in 1998 (Smith *et al.* 2003) revealed that the industry used 424 million tonnes of materials. Over 150 million tonnes of waste was produced (90 million tonnes generated on construction sites and 60 million tonnes by product manufacture). In addition, 30.1 tonnes of emissions were generated. The study concluded that the most significant barriers to achieving improvements in resource productivity were the fragmented nature of the construction industry and the absence of a body to be able to provide strategic direction for setting priorities and policies.

The Strategic Forum for Construction Logistics Group's report (2005) *Improving Construction Logistics* identifies the following factors preventing the construction industry from addressing logistics:

- The lack of an incentive to change.
- Construction projects seen as a one-off and therefore difficult to optimise logistics for long-term benefit, like the retail sector.
- The fragmented nature of the construction industry and lack of direct employment.
- Inadequate advance planning of projects and short lead times.

- The lack of cost transparency in the construction process hinders the identification of potential savings from improved logistics.
- Inadequate information flow.
- The lack of understanding, trust and confidence in supply chains.
- Clients (and others) believe that project costs already include for appropriate logistics resources to be committed to projects.

The Strategic Forum for Construction Logistics Group contends that logistics will not be adequately addressed until the construction industry works in a more integrated way, with all members of the supply chain. The Forum recommends an action plan based upon the utilisation of integrated project teams and supply chains and the use of off-site manufacture and modern methods of construction (MMC) and greater use to be made of information technology.

Cultural barriers and the level of price uncertainty amongst the variety of logistical activities inhibit the adoption of specialised logistics techniques. It is difficult to accurately predict the flow of materials through the supply chain and price the logistics involvement of each party. Therefore, contractors are reluctant to implement radical change, fearing that it might involve significant cost and act as a disadvantage in the fiercely competitive marketplace. In addition, it is possible that some members of the supply chain believe that their organisation benefits from operating inefficiently, a lack of transparency shields these members and it is difficult to prove and extract tangible benefits because of the level of price uncertainty.

The lack of certainty in predicting the flow of materials through the supply chain highlights the benefits of value stream mapping (VSM) when improving supply chain performance. VSM provides the opportunity to eliminate members of the supply chain which are unable to contribute sufficient 'value' to the product, and is discussed further in Section Three.

Figure 2.6 demonstrates the transition from a disjointed logistics approach to a controlled approach, where an increase in the control and improvement in handling and distribution of materials increases efficiency and reduces waste.

Cultural barriers to implementing integrated logistics in the construction industry

Before advancing the case for the implementation of integrated logistics management, it is useful to consider the barriers that currently impede its general acceptance in the UK property and construction industry. The barriers identified are inextricably linked with the industry's prevailing traditional and conservative culture. Much has been written about culture during the past decade and it is therefore relevant to appreciate its origins and development before applying it specifically to the property and construction industry.

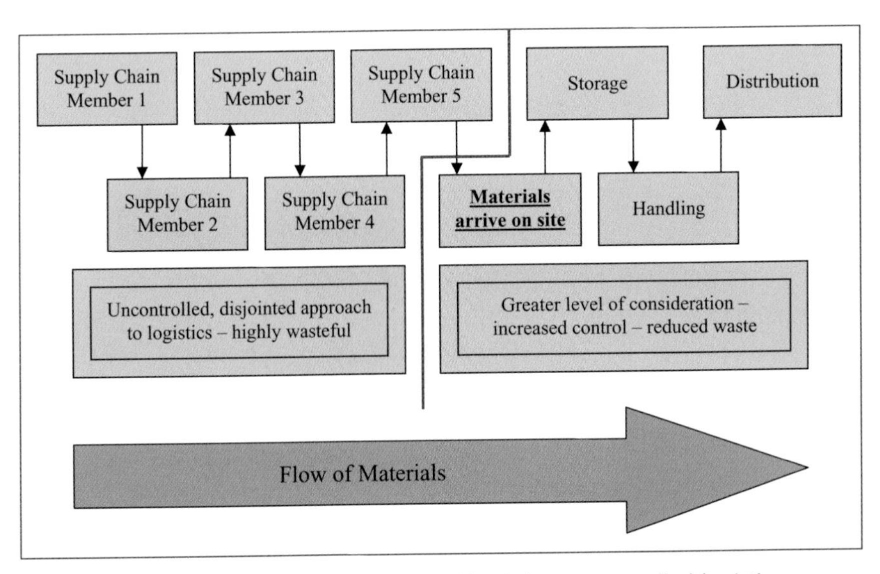

Figure 2.6 The transition from disjointed logistics to controlled logistics

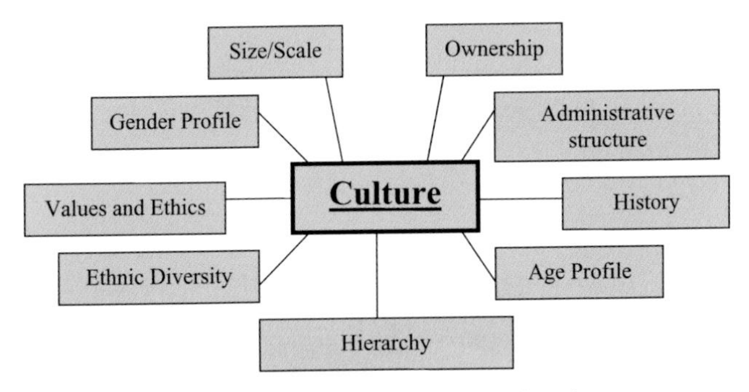

Figure 2.7 Factors influencing an organisation's culture

The use and meaning of the word 'culture' has developed and expanded significantly since its original horticultural derivation from the Latin verb *cultivare*, 'to cultivate'. The British anthropologist Edward Tylor is credited with being the first person to use 'culture' in an anthropological sense in 1871. Tylor defined culture as 'that complex whole which includes knowledge, beliefs, art, morals, law, custom and any other capabilities and habits acquired by man as a member of society'. Tylor's seminal nineteenth-century definition has formed the basis of many contemporary definitions applied in a business context, now recognised as 'organisational culture'. Buelens *et al.* (1999) suggest that culture is 'the set of shared, taken for granted implicit assumptions that a group holds and that determines how it perceives, thinks about and reacts to various environments'. Organisational culture influences behaviour at work and reflects the values shared amongst its members. Figure 2.7

illustrates the factors influencing an organisation's culture, which also affect its ability and willingness to embrace change.

Barthorpe (2002) suggests that the property and construction industry in the UK and some other developed countries exhibits a distinct, albeit complex, culture which comprises many interacting 'sub-cultures' and 'counter-cultures' that set it apart from any other industry and therefore make it unique.

To comprehend the underlying causes of the construction industry's inability to embrace improved logistics techniques in line with other industries it is necessary to appreciate the nature of its culture and identify the barriers which exist.

Resistance to change

Industry reviews by Latham (1994), Egan (1998), Bourn (2001) and the Strategic Forum for Construction (2002) have identified that many practices are outdated and are overdue for reform. The need to deliver projects on time has actually restricted innovation within the industry and has led to a tendency to stick to the tried and tested methods, according to Rogers (2004). Blayse and Manley (2004) suggest that this problem is compounded by the one-off, bespoke nature of most building projects, whilst Dubois and Gadde (2002) contend that the organisational structure of the project team, that is a temporary coalition of small firms to complete a building project, provides little incentive to research and develop new, improved methods of working.

Buelens et al. (1999) argue that the need to change in an increasingly competitive marketplace makes organisational change a necessity. Businesses are forced to embrace more effective methods of operating to satisfy consumer demand for greater value and lower prices. The concept that 'innovation is inseparable from progress' is particularly true of construction, with many projects delivered behind time and over budget (Robinson 2004).

Logistics constraints

Every construction site has a different set of constraints that affect construction operations, including logistics. The nature of the constraints will depend on a number of factors, such as the location of the site, the nature of the work and the working environment, the potential for construction activity to affect site neighbours and the social policy of the client. The contractors' ability to deliver a project will often depend upon their ability to devise and implement a sustainable construction strategy from the outset.

Logistics is the most significantly affected site-based activity. It is also one of the most important activities since workforce productivity levels depend upon the punctual delivery of mechanical equipment and materials, which ultimately affect the completion date of the project. The delivery of equipment and materials may be affected by factors on and off-site, for example:

- physical constraints such as low bridges and one-way systems etc. (off-site) and the lack of storage space or restricted access due to narrow corridors and existing structures etc. (on site)
- legislative constraints such as the Town and Country Planning Act (1990), Environmental Act (1995) and the Manual Handling Operations Regulations (1992)
- environmental and social constraints, e.g. where the construction strategy has to consider minimising noise, dust and disruption and consider the proximity of watercourses, neighbours and existing structures and roads etc.
- financial constraints which might restrict the deployment of mechanical resources etc. that a contractor is able to use on a project.

Among the most challenging environments for construction work to take place are those where construction activity has to proceed around the uninterrupted commercial needs of the client, often whilst being occupied by the public, for example during shop refurbishments, airport terminal extensions and hotel renovations. Significant logistical constraints and financial consequences are also encountered by contractors involved in motorway repairs, Tube station maintenance and railway line engineering works. In these situations, circumstances do not permit the setting-up and operation of a permanent construction site. This often means that there is no space allocated for materials storage and construction activity is restricted to specific pre-arranged and carefully planned shut-down periods. To compound this problem the client is likely to put greater emphasis on time certainty and speed of construction because of the high cost and level of importance placed on resuming normal business operations.

Security also acts as a major constraint on many projects. Government premises, military facilities, the public transport infrastructure and even major sporting venues remain a significant target for terrorists. Implementing a logistics strategy which offers a high level of control for security sensitive sites is therefore essential.

Logistics on sites located in major towns and cities is often severely restricted by the lack of storage space, congested roads and the fundamental need to minimise the disruption caused to other businesses and residents in the area surrounding the project.

Overcoming each of these constraints to achieve optimum project performance requires careful consideration and planning long before work starts on site. However, the traditional approach to logistics often does not offer a sufficient level of control to accomplish this and creates many of the problems identified in Figure 2.4.

References

Barthorpe S (2002) The origins and organisational perspectives of culture. In: Fellows RF and Seymour DE (eds), *Perspectives on Culture in Construction*, CIB Publication No. 275. International Council for Research and Innovation in Building and Construction, Rotterdam, pp. 7–24.

Blayse AM and Manley K (2004) Key influences on construction innovation. *Construction Innovation* **4**(3): 143–54.

Bourn J (2001) *Modernising Construction.* National Audit Office, The Comptroller and Auditor General, London.

Building Research Establishment (2003) *Construction Transport: The next big thing.* BRE, Watford.

Buelens M, Kinicki A and Kneitner R (1999) *Organisational Behaviour.* McGraw-Hill. London.

Chartered Institute of Logistics and Transport in the UK (2006) http://www.ciltuk.org.uk, [accessed 14th February 2006].

Considerate Constructors Scheme (2006) http://www.ccscheme.org.uk, [accessed 20th March 2006].

Dubois A and Gadde LE (2002) The construction industry as a loosely coupled system: Implications for productivity and innovation. *Construction Management and Innovation* **20**(7): 621–32.

Egan J (1998) *Rethinking Construction.* The Report of the Construction Task Force. HMSO, London.

Fairs M (2002) Logistics giants UPS and Exel to target construction. In: *Building* 28th June 2002, The Builder Group, London.

Latham M (1994) *Constructing the Team.* HMSO, London.

Robinson G (2004) *I'll Show Them Who's Boss: The six secrets of successful management.* BBC Books, London.

Rogers P (2004) *Construction Logistics Consolidation Centres conference speech*, Royal Aeronautical Society, 6th October 2004.

Smith RA, Kersey JR and Griffiths PJ (2003) *The Construction Industry Mass Balance: Resource use, wastes and emissions.* Viridis Report VR4 (Revised). Viridis, Berkshire.

Strategic Forum for Construction (2002) *Accelerating Change.* Rethinking Construction, London.

Strategic Forum for Construction Logistics Group (2005) *Improving Construction Logistics.* Construction Products Association, London.

Taylor D (1997) The analysis of logistics and supply chain management cases. In: *Global Cases in Logistics and Supply Chain Management.* International Thomson Business Press, London.

Tylor BE (1871) *Primitive Culture: Researches into the development of mythology, philosophy, religion, language, art and custom.* Murray, London.

Chapter 3
An Introduction to Practical Logistics

My logisticians are a humourless lot ... they know if my campaign fails, they are the ones I will slay first.

Alexander the Great

Logistical support for military manoeuvres is an ancient art, which great generals such as Alexander the Great understood to be crucial to the success or failure of their campaigns. Techniques remained largely static, however, from the ancient world until the advent of modern technology. The history of military logistics – centuries of unchanging practice followed by a step change – seems to provide an apt comparator for construction, where supply methods have remained relatively untouched by advances in other industries. Large numbers of people are amassed, as with the army, and while they do not fight an enemy, they do battle with the elements and the demands of the project. The construction industry's ability to build complex structures in difficult areas is extremely impressive, but currently their achievements are often despite, rather than because of, their approach to planning and supply. The modern discipline of logistics is a sophisticated part of many retail and manufacturing operations, but it has still to reach its full potential in terms of its utilisation by the construction industry.

It would be relatively easy to produce a plan with the level of sophistication commonplace in the retail or manufacturing industries for a construction project. Sadly, however, this beautifully detailed plan would be utterly unworkable, for a number of practical, cultural and contractual reasons.

This section is all about the right way to create a plan which is practical and workable on a real construction site. For this reason, it draws on practical experience more than learned sources. In years to come, construction may well have its own dedicated branch of logistics, which will create scholarly literature and technical process maps that future writers can draw on. For now, however, experience rather than learned authorities is, of necessity, the mainstay of this section.

Planning is everything

Logistics, like many other aspects of a construction project, have to be planned out in detail at the outset of a project if they are to run effectively and efficiently. Therefore, the preconstruction phase is vital.

This is the point at which practitioners need to develop an outline plan. In this plan, the logistics manager must formulate a strategy on how to approach the job at each stage of its development. Drawings and sketches are also vital. Perimeters, access points, footprints and vehicle movements, in particular, need to be considered carefully. Figure 3.1 shows how the placement of standing aircraft is taken into consideration on a construction plan of Heathrow Airport.

Context is everything. It's important to gain a thorough understanding of the local area. Who are the neighbours, and what are their requirements in terms of access and noise control? How does the local road network operate? What are the traffic levels? Some seemingly quiet roads become a rat run for commuters taking short cuts at rush hour. The logistics manager needs to understand these local constraints, and have

Figure 3.1 How the placement of standing aircraft is taken into consideration on a construction plan of Heathrow Airport

a detailed knowledge of anything likely to affect the project over perhaps a 300-metre radius from the site. For this reason, it's important to engage in extensive consultation with the relevant Local Authority in order to understand the road network. Consultations should also encompass the local community and its representatives – it's vital to understand their concerns, expectations and priorities. Fewings (2008) suggests that such exercises can hugely benefit the outcome of the project. Of course, the logistician isn't doing this in isolation. There should also be wide engagement with the views of the project team, to ensure that all issues relating to the management of the site are explored.

Then, the design of the building needs to be considered and understood. Data capture has to be methodical and extensive enough to give the logistics manager an adequate understanding of both the type of building and the method of its construction. One of the most important aspects to understand is the impact of any constraints. Constraints can take the form of the number of hours which can be worked, or of difficulties presented by materials, construction techniques or staffing levels. Labour histograms are a helpful way of visualising the workforce levels. Capacity analysis is also necessary, in order to establish detailed estimates of likely productivity levels, taking into account the number of unloading points, cranes and hoists, etc. Working out what can feasibly be achieved in a typical day, and the most effective way of deploying the trades, is a key exercise at this stage to establish spatial constraints and any likely impact on the execution of the design. It's incredible how often a design calls for, say, a large piece of steel when there isn't the capacity for an articulated lorry to deliver it to site. Technology has now given us the capacity to work out the notional spatial requirements for deliveries, and aspects such as turning circles can now be mapped out on a computer using swept path analysis. We have to be careful with these, however: computers can't take variables such as the driver's skill level into account. Some people are going to need far more space than produced by a software program in order to manoeuvre a large vehicle!

This is where gaining sufficient local knowledge is crucial, so that plans can be assessed and tested for their viability in practice. The logistics manager should consider what type of vehicles, cranes, hoists and manual-handling equipment (MHE) will be needed for unloading deliveries. This is known as 'common user plant'. To be effective, the logistics manager must understand the construction programme thoroughly, because it is sometimes necessary to provide input into purchasing decisions in order to make sure the project gets the right equipment. For instance, it can be very tempting for purchasers to acquire smaller hoists, thinking that they're the most cost-effective option. This looks great on paper, but isn't so helpful when it's found that the materials required won't fit them! It's vital to avoid these false economies to create a well-run project. On-site productivity can only proceed at the pace of the materials deliveries.

When planning for unloading operations, it's also important to consider the volume and flow of materials, and the capacity to unload them in the time available. This is discussed in depth in Chapter 5, but the key point to remember is that if you have 20 hours' unloading time and each delivery takes an hour don't plan for 25 deliveries. This sounds obvious when it's stated, but time and time again logistics operations come unstuck because of precisely this kind of mistake. The purpose of the logistician is to introduce simplicity, because complexity equals risk. Those who have to execute the plan should always be considered at the planning stages.

There are four main resources competing for access and egress: people, materials, plant/tools and waste. The logistics plan will be significantly linked to the site health and safety plan. Increasingly, security constraints are also having a significant impact on the functions of the logistics team.

Finally, the likely effect of the weather on timing and logistics is often not adequately accounted for at the planning stage. Many proposed building programmes in the UK tend to be written as if for a country which never sees rain, let alone more serious weather conditions. In reality, the logistics manager needs to be aware that their plans can be severely disrupted by these unpredictable forces of nature. The undesirable combination of wind and tower cranes is one of the most obvious examples of this, but there are many other potential issues. Storms, lightning and rain can all have a huge effect on groundworks, for instance. Also, there can be damage to many elements of a building until it is waterproof. Even extreme heat can have an undesirable effect, particularly through the warping of materials.

Waste not, want not?

At the other end of the materials lifecycle, the logistics manager needs to be aware of waste generation issues. The issues involved in this area can be extremely complex.

It is well documented that the construction industry generates a substantial proportion of all waste outputs, amounting to between two and five times the quantity of household waste in European countries (Nowak *et al.* 2009). Whilst the pressure to be seen to be green affects all industries, it is therefore becoming a particularly pressing issue for construction firms. However, existing practices and typical contractual arrangements, it will be argued, often perpetuate wasteful approaches to materials usage.

To begin with, waste removal services are generally arranged using a lump sum contract. This means that the provider has to become highly skilled in guessing how much waste will be generated by a project. If this becomes part of the logistics manager's remit, it needs to be understood that full information provision is crucial to the creation of an equitable contract which benefits both parties.

There is also a level of uncertainty about what, precisely, constitutes waste. This matters because a waste carrier's licence is required to transport waste. It is, however, unclear whether virgin materials which leave site unused constitute waste. The issue of waste reuse and recycling is explored in more detail in Chapter 10.

What does seem clear from observation on site, however, is that contractual arrangements, such as pricework, where payment is linked to output, provide a strong disincentive for subcontractors to spend time minimising or organising their waste. Trade contractors are rewarded for speed, rather than for environmental stewardship of materials, so the temptation for the contractor is to discard the packaging *in situ* and crack on. This pressure to work quickly often means that it's hard to get a trade contractor to put anything in a waste bin, and getting the right waste in the right bin is even harder! This is a jocular observation, but the more serious truth behind it is that the contractual conditions create no incentive for them to use their time differently.

Despite these contractual disincentives, there is undoubted pressure to present a company as environmentally friendly. This pressure to demonstrate a good green performance can lead to a way of reporting statistics that does not necessarily accurately reflect actual project performance. Waste contractors commonly arrange for waste to be transported to transfer stations, which separate waste for recycling. However, the transfer station is only able to report aggregated figures on the percentage of recycling achieved from across its waste intake, which will come from a wide variety of sources. The waste contractor will nevertheless get a certificate saying that a certain percentage of waste (say, perhaps, 80%) was recycled, and his client will be delighted to be passed this information, which may well be used as a key indicator of recycling performance for the project. In reality, however, the aggregated figure from the transfer station and the amount of recyclable material from the particular project may bear very little relationship to each other.

Alternatively, flagship projects may manage the pressure to achieve green performance targets by creating banner targets – such as, say, 50% of materials to be delivered by rail and water. These sound impressive, but if the target percentage is interpreted by weight then the target can easily be achieved through the supply of aggregates – hardly a radical departure from normal practice.

These observations may sound somewhat negative, but there is a good reason for exploring these issues. Due to the importance of the issues driving the sustainability agenda, which are seen by many learned commentators as the gravest threat ever faced by the human race, there is an unprecedented pressure on companies to do, and to be seen to be doing, the right thing. At the same time, a reliance on questionable statistics and a reluctance to initiate a fundamental reform of both cultural and contractual norms (remembering how closely these two norms are actually entwined) militates against the industry achieving optimal performance.

However, these issues shouldn't be allowed to obscure the many positive opportunities offered by the sustainability agenda. By taking a proactive and efficient approach to waste management, a contractor can achieve significant cost savings whilst benefiting the environment. Green company policies are undoubtedly beneficial for a company's image, but a real commitment to achieving a better environmental performance can also have an extremely positive effect on staff morale, materials efficiency and the business bottom line.

Communication is the key

Much of logistics management is about taking actions that might be taking place already on an ad hoc basis, giving them a name and assigning responsibility for them. Ultimately, the discipline of orderly logistics needs to become part of the organisational culture, but thoughtful communication is vital to getting it embedded.

Communication is, however, something that has to be approached with care. Emmitt and Gorse (2003) note, 'Construction professionals enter into communication with diverse perceptions, attitudes and values.' The same commentators continue, 'Words can have very different meanings between fields of specialization, yet people do not usually define the words that they are using.' Even the word 'logistics' is, in itself, ubiquitous to an extent that it can mean many different things to different people. Therefore, if you are invited to manage logistics, it's important to establish at the outset exactly what you're being asked to do. Are you being asked to plan the site establishment or manage materials flow, or both?

There are several sophisticated tools around logistics, and like most specialities, it has its own jargon, leading the professional logistician to get involved with such strange-sounding tasks as developing a kanban system, integrating demand fulfilment or looking for inter-transit visibility. Despite that, however, the core job for a logistician can be described in extremely simple terms: in essence, their job is to get basic processes functioning consistently.

The secret to achieving this is, once again, good communications. In logistics, communications have to be tackled holistically. Written and oral communications need to be timely and clear, but physical communications (the movement of people, materials, tools, plant and equipment) also need to be addressed. In these days of ubiquitous information communication technology (ICT), we can sometimes forget that roads are as much part of communication as satellites.

For that reason, the logistics manager needs to be able to communicate at several levels. On the one hand, for instance, there is a requirement to be able to engage with the technicalities of ICT tools such as a delivery management system but, on the other hand, effective communication with truculent delivery drivers is equally essential to the job.

The logistics manager should be invited to site progress meetings and directors meetings. They need to be kept informed of progress on site so that they can look ahead to requirements for the next week or month of productivity. However, these meetings cannot be relied on in isolation to outline requirements. The logistician also needs to be in constant touch with the reality of progress on site – this can sometimes provide a different perspective from official reports that are discussed in formal meetings! The logistics manager therefore needs a kind of intelligence-gathering system. Traffic marshals and supervisory staff should be encouraged to keep feeding information up the line to help inform an understanding of what is likely to happen in reality.

It is a noted characteristic of the industry that it tends to over-promise and under-deliver. In fact, this is a fairly universal (and very human) reaction to uncertainty on projects. Whenever there is a need to estimate something, there is a temptation to draw on corporate objectives (for example that task X will be done in three weeks) rather than experience (for example that task X usually takes six weeks, and longer if a key staff member is away) in formulating the figures. This refusal to face up to uncertainties has been described by Chapman *et al.* (2006) as a 'conspiracy of optimism', and it has been argued that this commonly compromises the accuracy of estimates.

This is especially true in the area of deliveries, where there is a constant temptation to repeat the promises of others, without any ability to establish the veracity of the reports. To mitigate this, it's important to develop the habit of 'managing by walking about'. The logistics manager should always work to establish a good relationship with members of the key trades, so they can make informal enquiries to establish the true state of play. There's a need to keep evaluating sometimes conflicting information. Otherwise, there's a danger that the plan for the site's logistics will be based around a false premise of expected activity. This usually leads to the wrong resources in the wrong place at the wrong time. To avoid this, both formal and informal information sources need to be constantly evaluated and cross-checked, with a view to establishing the cause of any discrepancies as quickly as possible.

In the First World War, when communication technologies were in their infancy, messages often became distorted by a process not dissimilar to Chinese whispers, leading to the famous miscommunication where the request to 'send reinforcements – we're going to advance' got mangled on its way back to headquarters into 'send three- and fourpence – we're going to a dance'.

This is a valuable reminder that effective communication has two components. We don't just need to be able to present information effectively, we also need to listen, and question where necessary to establish that we understand what we've heard.

Language issues also need to be taken into account. English is an extremely idiomatic language, and therefore we need to think carefully about how we communicate with workers with only basic or intermediate English language skills. Otherwise, we risk similar incidents to the one occasioned when an Eastern European carpenter was told that a project needed a zebra crossing. The project manager returned to find the silhouette of a zebra, carefully created out of ply board. If we think about this incident, we quickly realise that the worker had done exactly what was asked – the zebra had been affixed to the road, and so was clearly crossing it – the problem lay with the instructions.

Good communication requires adequate information, and in the complex world of construction this can sometimes be a challenge. There can be cultural or contractual reasons for withholding information, but a good logistics manager will work to overcome this where possible. Of course, contractual requirements can't always be surmounted, but any learnt disinclination to share information, perhaps because 'knowledge is power', needs to be challenged. Logistics also requires a sophisticated understanding of the interdependencies of the various relationships, and to assist in creating team coherence by managing up and down as well as across and externally.

A key role for the logistician as communicator, especially on sizable projects, can be to mitigate some of the effects of package management on the delivery of construction projects. This approach to management can sometimes lead to a 'silo mentality'. Naturally, the project manager will have an overview across the entire project, but their role necessitates that they spend much of their time focused on construction productivity activity and meeting programme dates.

The overall operational focus needed to bring together the people, material, plant and equipment in the right place at the right time can sometimes be missing from this approach to working. Therefore, if the logistics manager can provide this overarching perspective, accounting for all activity on site and some activity off-site, they can make a valuable contribution to efficiency. For this to work, the logistics manager needs to be given enough authority to implement the effective coordination of tasks.

As previously noted, a complex building project has some similarities with the army, and the logistics function can be rather like the military 'command and control' approach, moving personnel and machinery in concert to their most effective position. This can reap benefits for a project, but if this system is not adopted there is a danger for the logistician that they are left with the responsibility for such manoeuvres, without the necessary authority to direct them. Obviously, there will be times when both package managers and logistics managers will need to refer to the overall authority of the project manager, but the author's experience is that if multiple package managers override the logistician, who needs to be given a relatively free hand to build a cohesive operational site plan, things can happen in a way which is both irrational and expensive to the project.

Because of the commercial drivers for the trades delivering various elements of the packages (payments by results does not generally account for situations where the 'results' stop another trade from doing their job), there is normally no incentive for the workers to liaise and consult with others on the site. After all, if they have to remove and rebuild their work, they stand a good chance of being paid twice! This can lead to such risible situations as pre-assembled doors needing to be dismantled and rebuilt to fit through apertures in walls that did not exist when the delivery was planned the day before. The best way to prevent this, and ensure that the right communication and liaison takes place, is to give the logistics manager, who is perhaps best placed to see the bigger picture from an operational perspective, the authority to control such interfaces.

This is perhaps the place to mention knowledge management, a growing discipline which has long been advocating the need for organisations to capture information efficiently in order to maintain their competitive advantage. Harris and McCaffer (2001) observe that, despite the complexity and knowledge-based nature of construction, there tends not to be many board-level appointments of knowledge management professionals within construction firms. Nevertheless, awareness of its potential advantages is growing and information sharing undoubtedly has a major role to play in helping companies to achieve their strategic goals.

However, it's also important to remember that sharing even the most prosaic information needs to be a formalised part of everyday working life in order to avoid the kind of disastrous mistakes that are perhaps best summed up by a verse from Lewis Carroll's *Hunting of the Snark*:

He had forty-two boxes, all carefully packed,
With his name painted clearly on each:
But, since he omitted to mention the fact,
They were all left behind on the beach.

Standardisation and pre-assembly in construction

Standardisation can be a very emotive issue in construction. It's often said, and not without truth, that it's a logistician's dream, but an architect's nightmare. Within the hierarchy, architects' preferences tend to carry more weight; considering that they are responsible for the creative vision of the project, this is how it should be. However, in elements of the building where artistic flair is not so essential – such as internal services, for instance – more utilisation of standardised products could help create a step-change in productivity.

When the consolidation centre at Heathrow was set up, the contractors were asked to submit a list of all the parts or products they would be using. They said it would be impossible, because there were so many. When the logisticians insisted that they sit down with the trade contractors and go through their product list, most found that they were using, at most, around 80 products. Some would be using more, but 150 invariably marked the upper limit in terms of component variation. It's not always as complicated as we think it is.

Exercises such as these are invaluable to the logistics manager, because they give the level of detailed information which is so vital to all robust plans. When we know what we need for a project, we can start managing it effectively. For the same reason, logisticians can derive great assistance from the project's bill of quantities – if they are lucky enough for the project to have one! Since they provide a complete list of all the components, accompanied by the volumes required of each, to create a building, they are naturally a major point of reference for the logistics operation. If the project is being built without a bill of quantities, the logistician has to draw on the construction experience of the project team to advise on likely requirements and take an educated guess.

One of the problems in construction, of course, is that the industry doesn't always use the same name for the same bit of kit. There are different ways of describing different types of sand, which seems fair enough, but there are also different ways of describing the same type of sand! Orders for materials come in a mixture of metric and imperial – a request for three metres of 'four by two' is commonplace – and culture, naturally, has a part to play here.

Each trade has its own language; there is no formalised naming system for every component. The army has had a unique ID number for everything from socks to battle tanks since the Boer war, but the construction industry remains wedded to a system which, by its very nature, is prone to cause confusion. Whilst firms such as B&Q have the ability to barcode every single product, it currently remains a distant ambition for construction to embrace this technology.

The core problem is the tendency for architects to describe specifications rather than name products. This, as with so much in construction, is a cultural practice embedded in and perpetuated by the contractual allocation of risk. Trade contractors, therefore, are procuring materials by specification rather than by brand. The reason for this is obvious: if you specify a particular product and it fails, there are potentially serious contractual repercussions. If you specify using performance criteria instead of products, on the other hand, you are automatically mitigating that risk. The risk gets passed on to the trade contractor, who must therefore make the specific purchasing choice. Clearly, this system will not sit well with standardisation.

This kind of issue makes construction so hard to understand for logistics managers coming from a retail or manufacturing background. On visiting a construction logistics operation, their first questions will be to find out why we're not using radio-frequency identification (RFID) tagging or barcoding on the stock. The answer is that the reality of life in construction usually involves buying specifications rather than particular products. It's somewhat different in house-building, where, for instance, specific brick products are required for their colour, but in commercial building, the architect is king. (There have been rumours that the client is king, but the architect does not appear to have been informed, as yet.)

In short, whilst standardisation is the logistician's dream, we accept that there are currently significant cultural and contractual hurdles before this becomes a more prevalent practice. There are good reasons for adopting standardisation, but the status quo makes it difficult. The greatest incentive for designers, arguably, would be if they were fully appraised of the degree of improvement in the flow of materials to site. Then they would see the resultant improvement in efficiency, productivity and completion times that the wider implementation of standardisation could effect.

Off-site manufacture and pre-assembly techniques, however, are not generally subject to the same contractual impediments. Off-site manufacture and pre-assembly can therefore often be deployed as part of a robust logistics plan and should clearly be used wherever possible. It makes good commercial and operational sense to put components together in a safe, weatherproof and planned environment, rather than on a busy, congested and sometimes exposed site. If activities such as cutting and shaping don't have to take place on site, not only is congestion eased: it stands to reason that productivity and quality will improve also.

Managing this operation is a specialist function requiring an experienced logistician, since to fully realise the benefits there needs to be a streamlined operation to manage the way the materials are then packaged and distributed to the site. Attention has to be given to the optimal way of distributing these assembled components around the site, to the greatest benefit both to the trade contractors involved and to the project overall. The logistics manager will not be involved in the manufacturing process per se but will need to take full charge of handling all materials from the factory gate to the point of fix.

Despite the fact that the contractual inhibitors to standardisation don't generally apply to off-site manufacture and pre-assembly, there still appears to be some resistance to employing off-site techniques within the industry. There is also a danger that when trade contractors are tasked with managing off-site manufacture in-house they sometimes fail to realise the full benefits which can accrue from adopting manufacturing techniques. Off-site manufacturing can be a great aid to logistics on a construction site, but it really does need to be a manufacturing-driven solution.

Creating professional logistics operations in the construction industry

Logistics is all about making things happen in the real world. To get it right, therefore, it's important to take a realistic view of all facets of the operational context.

It is often observed that productivity and performance on construction projects suffer because of a fragmented supply chain. Indeed, as far back as the *Simon Report* of 1944, the industry has been characterised as suffering because of the contractual models which reinforce the complexity of relationships on site (Murray and Langford 2003). However, despite successive government reports and interventions, the problem remains, perhaps because, as London (2007) argues, many models for integration ignore the reality of 'a sea of firms of buyers or sellers in a variety of competitive markets exchanging commodities'. There are undoubtedly both advantages and substantial barriers to greater integration, and the logistics manager needs to have a realistic appreciation of the issues that this creates.

Holistic engagement with all facets of a project can be more of a utopian dream than a practical prospect. Therefore, the logistician needs to work within the parameters of what is possible, having determined as accurately as they can precisely where these parameters lie, and create strategies to mitigate any problems created by the complex relationships between different members of the team. In this way, a professional approach to logistics can actually become a key strand in the drive to achieve the kind of integrated project team recommended in *Rethinking Construction* (Egan 1998). This report noted that fragmentation underpins an adversarial culture. By solving problems early and smoothing the path to integrated working, logistics can help mitigate the risk of difficult outcomes on complex projects.

A closely related issue which needs to be understood is that of culture. The culture of the construction industry is arguably unique, and both the positive and negative aspects of the status quo need to be understood. The author of this section, a leading logistics practitioner with a military background, has (perhaps surprisingly) found that construction organisations can be more autocratic than the army! There can be a tendency for executive decisions to go unchallenged, but a good logistics manager needs to make robust interventions for the good of the project when it becomes apparent that the plan on paper is not going to translate to success in practice.

Whilst construction brings its own challenges as a field of operations, it's nevertheless one of the most exciting and challenging industries to work in. Its resourcefulness and creativity is brought to bear daily to create bespoke solutions for unique designs, and the ability of project teams to bring seemingly impossible situations through to timely completion can seem little short of miraculous. Of course, it is possible to bring a project to completion without employing a specialist logistics manager. Nonetheless, as is demonstrated in the conclusion to this book, there is a compelling financial argument for professional logistics in construction. Good logistical planning reduces reactive 'fire-fighting', which can be a substantial drain on project resources. Specialist logistics managers can therefore mitigate risk and reduce costs by increasing certainty, productivity and predictability within the schedule.

Due to the complexity of construction projects, the author believes that we have to be careful when we look to manufacturing for solutions. The motor industry's solutions for efficiency are sometimes seen as a panacea, but the challenges they have to overcome are very different. For instance, whilst one car manufacturer offers some 240,000 variations to customers, many of the options are around small details such as seatbelt colour, which can be customised quite simply. In construction, on the other hand, changes can have a dramatic effect on the following trades. The complexity of the project (and the impact of changes) is therefore, it is argued, quite distinct from those of manufacturing industries.

To manage logistics in such a complex environment, you have to plan for contingencies A, B, C, D, E and sometimes F, bearing in mind that the outcome might yet be G or H! When we consider that on some projects construction can begin with as little as 20% of the design complete, the need to have contingency plans to consider the implications of design change becomes immediately apparent. There's a military saying that 'no plan survives the first contact with the enemy'. War is organised chaos, and construction is not so very different. Therefore, whilst planning is absolutely essential, the logistics manager has to be flexible and agile, updating plans constantly to respond to changing requirements on the ground.

Good logistics is about keeping a holistic focus on practical challenges, in a situation where perhaps dozens of trade contractors are focused on delivering their own work package and managing their own risks. The key to increasing efficiency is in successfully fostering collaboration between all the different actors in a project and in creating integrated controls. Not only do the main contractor and all the trade contractors need to communicate with each other: to achieve optimal performance, a good logistics plan should allow for collaboration between all the interested parties, from client to designer to the local community, and any other stakeholders with an interest in the running and outcome of the project.

The logistics manager must also provide a lead in solving a number of practical issues, such as the need to create a watertight building, segregate waste and provide adequate temporary accommodation. Much of these tasks were historically seen as within the remit of the site manager, but due to the complex nature of construction contracts they are now often fully occupied with issues of quality, risk and cost allocation, leaving the logistical challenges of coordinating the work to the logistics manager.

This is as it should be, particularly if the project has the input of a trained logistics specialist. This is a far better solution than leaving everyone to solve their own logistical problems. After all, a logistics expert wouldn't be expected to fix the lighting, so why expect electricians to be expert in logistics?

Employing a logistics specialist is therefore desirable, but at the time of writing it is likely to entail looking outside of the construction sector. Construction companies generally do not currently train home-grown logistics managers, so logisticians working in construction have usually been brought in from other industries or from the military. As construction logistics becomes an increasingly specialised and professional function, however, it seems likely that some knowledge of the discipline will soon come to form an important part of the training of any would-be construction or project manager. The construction industry, whilst generally perceived to be an extremely traditional sector, is capable of making very impressive leaps forward in some aspects of its performance. Whilst there is still some way to go, the improvements in safety culture and in attendant attitudes has been substantial, within just twenty years. It seems likely that a similar attitude shift might take place towards logistics.

However, a good logistics manager will not be looking to revolutionise the approach to the project, but rather to improve overall efficiency by formalising the best of what goes on ordinarily into a standardised procedure. Figure 3.2 illustrates the logistics cycle.

Given the complexity of the task, how does one go about forming a coherent and effective plan? First, it has to be informed by company

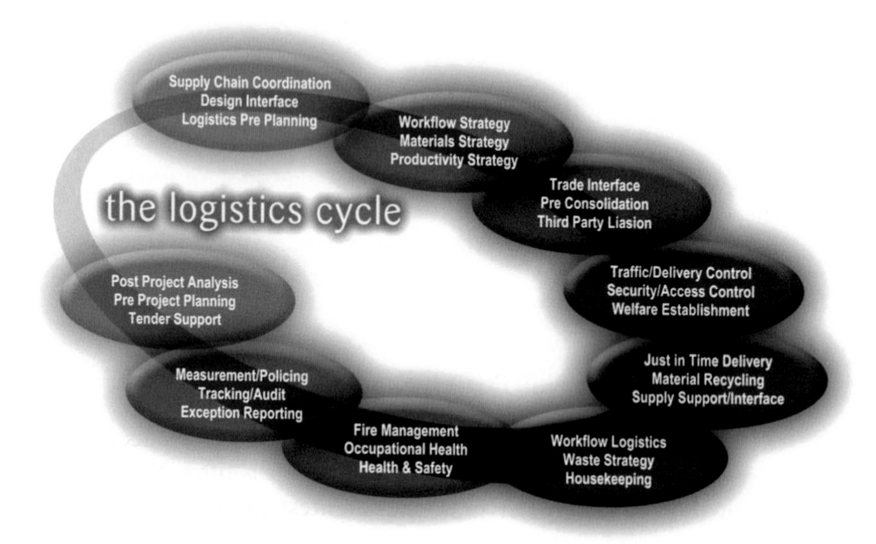

Figure 3.2 The logistics cycle

policy. It should be rooted in standard operating procedures, but the top-level company strategy should also be referred to in order to inform the overarching aim of the plan. Although much of the work of a logistics manager – as will be reflected in the detail of a good plan – is about dealing with minutiae, it's important not to lose sight of the project's goals and their strategic importance.

Logistics is not about reinventing the wheel. After all, the discipline was brought about as a result of the wheel, which was a key technology for ancient invading armies! However, the industry needs to, if not reinvent the wheel, at least break the mould, since there are huge advances to be made in terms of performance. Much of the methodology for creating a good logistical function is not rocket science, but it does involve recognising and coordinating a large number of otherwise disparate functions. Logistics plans do need to be linear, to reflect and facilitate the overall programme, but they need to be subdivided into task groups as well.

Site logistics plans (also known as 'external' or 'site-wide plans') will include provisions for cranes, hoists, accommodation, canteens and people access, whilst the internal logistics plan will include housekeeping, traffic plans, pedestrian segregation, temporary services, wheel wash, water and rain management and signage. This internal plan is, in effect, preliminaries (prelims) management.

The way that materials flow in, and the outflow of waste and materials for reuse or recycling, is a completely separate plan, but it feeds into and shapes both the external (site-wide) and the internal (prelims) plan. Materials flow is the life-blood of any working site, and therefore needs to flow as consistently as a healthy heartbeat! Planning the materials flow is a complex and often frustrating task as the pertinent information can be difficult to obtain.

Logistics managers who streamline these operations effectively can have a valuable role to play in ensuring that organisations reach their strategic goals. Paradoxically, however, they are also responsible for some of the most mundane functions on any construction site, and this responsibility can bring pressures and challenges of its own. As well as being a strategist and master tactician, prosaic tasks such as manning the security gate and organising breakfast fall under the logistics remit. If there is any kind of hitch, particularly when the task is a seemingly simple one such as serving breakfast, the feedback is likely to be immediate and scathing. Perceived failure can be judged especially harshly due to the mundane nature of such tasks, although in reality the challenges of feeding several hundred people in a timely and efficient manner can be far from simple. When these functions are operating well, however, no one will notice. That is often the acid test for success in logistics! Particularly, because the logistics operation is often the 'front of house' team that visitors meet first, the slightest non-compliance in terms of personal protective equipment (PPE) or adherence to process will be picked up immediately. If a visitor to the project is a VIP, this can be particularly unfortunate, but whoever is at the gate the team must remember that, to a very great extent, perception is reality. To cite another truism, first impressions matter. The importance of the utmost professionalism and accuracy, therefore, as well as rigid adherence to detail, must be constantly stressed to the team.

A logistics specialist also has to be conscious of the limits of their role. The logistics team is there to undertake the overarching logistics. Their job is not to fill a scope gap or to act as the 'emergency services' for subcontractors with failed logistics plans. Where a trade contractor has failed to make their contribution to site housekeeping, for instance, it's important to ensure that the logistics contractor is not automatically expected to deploy their own resources to compensate for them unless this formed part of the original tender. If it happened only occasionally, it would not be an issue. If, however, the logistics team is doing significant amounts of extra work on a daily basis, that would need to be formally renegotiated.

Small sites

Finally, whilst there is a tendency to associate specialist functions like logistics with large and complex projects, proper logistics management is perhaps even more vital to smaller projects, where there are likely to be limitations on both resources and space. Therefore, all the techniques discussed in this section can be adapted to improve efficiency on even the smallest sites. Whilst such projects might not warrant a specialist logistics manager, good logistics management should be a key part of professional practice for all operational managers in the field. Even a dedicated logistics professional will sometimes need to buy in specialist support to ensure optimum planning for some of the most technical tasks, but many construction managers could enhance their professional practice by developing a good understanding of the basic functions and benefits of a systematic approach to logistics.

Professional logistics solutions for construction are already starting to seem more attractive to many large contractors. This may be due to the complexity of the project, the aspirations captured in the company's corporate social responsibility (CSR) policy or a need to compete on productivity targets. Although there is far from being a consensus in the industry regarding the benefits, there is nevertheless a growing constituency of those who will embrace it. After all, if there is a bigger budget, it's an easier decision. Can a small project afford to implement professional logistics at all?

In the author's belief, the answer to that question is a definite yes. The logistician (or junior manager) working on a small site should take what's written in this book as a philosophy rather than a structure or a process. Then it can be applied to a £2 million shed as much as to a

£100 million commercial office building. The scale will obviously be different, but the principles are the same.

The beauty of many of the techniques discussed in this section is that they can be implemented in a very low-tech way. Mainstream logistics thrives on very sophisticated models and processes, which meet the needs of modern manufacturing and retail. They are servicing a very controlled environment, which is entirely focused on quality and productivity. Logistics as adapted for all sizes of construction project, by contrast, is about using a much more flexible approach to streamlining an innately complex environment, and so it relies less on technology and more on basic process management. Therefore, it's far easier to scale the solutions down to smaller projects. Any enterprising junior manager could adapt many of the ideas in this book.

Even the smallest domestic project could benefit from improved logistics. Not only would extensions and patios get built more quickly and efficiently, the entire industry would benefit from the resultant improvement in the perceived performance of the builders themselves. Ultimately, the entire industry is often judged on the basis of the public's own experiences as micro-clients. What do they see? Too often, they see bricklayers waiting around half the morning for a delivery. They see huge surpluses and waste. The result is that the whole of construction is seen as an industry of tea-drinkers. Nine times out of ten, the person actually paying the bill may not be able to analyse or articulate the problem with paying for skipped new materials on a cost-plus basis. Nonetheless, they still suspect that the whole process has been suboptimal in terms of efficiency and value for money, and they may well be right. This has, it is argued, helped to reinforce an unfairly derisory attitude to builders. Staged deliveries, efficient management and one nominated person with responsibility for managing logistics could improve both profits and perceptions on even the smallest site. It would be almost free to implement, and, done correctly, it could make local builders who created good supplier relationships almost unbeatable on price.

Much the same goes for modest-sized commercial sites. There will probably be a certain amount of effort put into logistical issues such as traffic management, but the idea of creating just-in-time deliveries to the workface is usually no more likely to happen here than it is on the average home extension.

Materials ordering needs forward planning rather than fire-fighting, whatever the size of the job. The major builders' merchants do little, at the time of writing, in terms of JIT, much less consolidated work packs. As markets tighten, however, even small builders might be able to use their buying power to persuade them to do it differently.

Another good reason for good logistics on smaller projects is that it can make a huge contribution to compliance with a raft of different requirements. Most of the following chapters in this section deal directly with at least one dimension of current legislation, be it fire safety, manual handling or the disposal of waste. All businesses today are operating in an environment where legislation is generally on the increase, and penalties tend to be getting more stringent.

It's also noticeable that there are increasing prosecutions of small businesses over issues which might once have been informally tolerated. Where there was once a tendency to focus on enforcing compliance within corporations, now there is an expectation that everyone will comply with basic standards.

These are all good reasons for applying the principles for professional logistics to smaller projects, but the most compelling reason of all for the small business is that it can save significant sums of money for little or no investment. The key message is not to get hung up on excessive documentation or learning specialist lingo. The general principles of construction logistics can be applied on any scale. All it takes is a little bit of forethought and planning.

Practical completion

'Practical completion', despite its contractual importance, is a term that is hard to define. Lord Justice Salmon, ruling in the *J. Jarvis and Sons* v. *Westminster Corporation* case of 1978, defined it as 'completion for the purpose of allowing employers to take possession of the works and use them as intended', rather than 'completion down to the last detail' (Ramsey 2007). In short, the building is ready to be occupied, and practically – although not quite – finished.

The practical impact on the logistician is that it can change the person in charge of the project, and payments, from the main contractor to the client. The client becomes the de facto principal contractor.

The other impact of practical completion is that it provides a major milestone in the life of the project. As the target date approaches, the speed at which work happens on site increases significantly. Obviously, in an ideal world, this date would be no different from a normal day, as tasks progress smoothly to completion. In practice, there is often a barnstorming finish, with all hands to the pump.

This race to the line is essentially a team sport, and all trades on site will be participating. For the logistics contractor, the challenge will be to manage a significantly accelerated requirement for materials. Discipline around deliveries will probably go out the window. No one is interested in JIT solutions. There will be an instinctive desire to get whatever's needed physically onto site, ready for the moment it's needed. People want to be able to see it and touch it, to reassure themselves that they will, against all odds, be able to get everything in place on the all-important date for practical completion.

If practical completion is not reached in time, the non-completion penalties can be pretty stiff. The pressure is palpable. This is the time for the logistician to remember Kipling's classic advice: 'If you can keep your head when all about you are losing theirs and blaming it on you'. Despite everything, the logistics manager needs to continue to focus on regulating the flow of material onto the site.

This will have to be achieved within a fiercely competitive environment. Not the least of the attendant challenges will be the fact that the plant is being stripped out and taken away. By this stage, it is likely that the cranes are gone and the hoists going. You would hope that the lifts are in by then, but this is not a certainty. You would also hope that by this point all the big items are already in the building, so use of goods lift and or passenger lifts will be minimal. Lifts will have to be heavily protected against damage. The reality, however, may be different. At this stage, the logistics team will be taking away significant quantities of waste and unused material, as well as the common user plant. Mobile elevating work platforms (MEWPs), it will be quickly discovered, won't fit in the lift. In any case, they will be needed later on, because the ceilings will still need to be cleaned. There is also the consideration that there will be a reluctance to use the lifts due to concern about invalidating their warranties. Much reliance will likely be placed on mobile cranes.

Throughout this chapter, there have been analogies with military logistics. At this point, it seems apposite to note that if normal construction is, in some ways, like a traditional land war, the drive to practical completion is more like fighting insurgents. Chaos must be constantly contained. No one, at any given moment, has a complete picture of what is going on.

On top of all this frenzied construction activity, the logistics manager will be trying to protect and clean detritus off finished products, and protect areas most liable to damage. Lifts are the obvious example, but areas such as computer rooms will need careful attention if they are to be guarded from damage successfully. The team will be starting to put the high-value items into the building, including finished products such as taps. Even minor damage, either accidental or deliberate vandalism, such as broken tiles can be a real setback at this stage. This is another argument in favour of having a fully skilled logistics team, who will have the loyalty and skill to avoid as many problems as possible. Agency staffing, in contrast, is something of a lottery. Workers who are living from hand to mouth have little reason to exercise the necessary care at this vital stage of the project.

As valuable as the logistics team undoubtedly is, its manager must be aware that it may be necessary to protect the team from a seven-day working week at this stage. Whatever the aspirations of the client, the legal safeguards on working weeks articulated in the Working Time Directive must be remembered. A five-and-a-half-day working week, with Sunday worked on a rota basis, is preferable. Of course, the more the industry embraces the benefits of professional logistics, the more the sprint finish at the end of projects could be avoided.

Even on the best-regulated projects, however, there are extra tasks which will take place at this stage. Cladding will be being fixed at the last minute, and water will be pushed through the permanent pipe system for the first time. Any damage to surfaces has to be remedied. At the same time, the team will be looking at the external works, and introducing landscaping where appropriate. Footpaths will need to be reinstated.

More waste will have to be removed. It's astonishing, even when a project has been well run and appears pretty clean, as the logistics team has been taking waste out twice a day, how much waste will suddenly appear. As you inspect hidden areas such as cupboards, you will likely find – quite literally – several tonnes more rubbish. Even on the best of jobs, using premier league contractors, you can still find yourself with truckloads of unused prime material to dispose of. This would normally have gone into a skip. The trade contractors just don't have the facilities to store this material, and so it's easier to treat it as waste. The logistician will make every effort to find better homes for this material, but it is almost inevitable that some will end up in landfill.

Importantly, the final clean will also have to take place. The technique for what is known as 'the builder's clean', which will already have happened, is to start at the top of the construction and work down, removing dust and grime on the way. This removes much of the dirt, but construction will still be going on. The final clean will remove the final debris, literally days or hours before practical completion and the all-important inspection walk round. There are a lot of politics involved in this. If the client is in a desperate hurry to accept the building, there will be a certain amount of compromise on details. In other circumstances, they may be as pedantic as possible. By managing the appearance of the building, logistics plays a huge part in the outcome of this exercise: it doesn't take a genius to work out that if a job is clean and free of waste, it will make a huge difference to the impression in creates.

The principal contractor will be aware of this, and, as the deadline approaches, they will occasionally issue an edict that every item of material and plant must be moved within 24 hours, or else it will be put in a skip. This can lead to the scrapping of a huge amount of plant and prime materials. In the author's opinion, this is a regrettable and unnecessary waste. It could be all avoided via a planned and orderly withdrawal.

The sprint finish, by contrast, often means that you end up having to remove all the facilities which would help the team complete the job. There will also be an increased need for people as mechanical aids disappear. For instance, there will be more guards, as there are no longer alarms. At the same time, the logistician will be making every effort to ascertain the legal owners of various items of plant. Sometimes, different trades use the same hire contractor, making it very difficult to work out who hired a certain item. There will be phones and radios to locate.

Some of these will be lost, and payment will have to be made to their owner. It is likely that they will have been lost for some time, although the hire charge will have been paid regardless. Mundane items such as fire extinguishers, pallet trucks and temporary fences will all need a home. If there is a buoyant market and new projects are starting, someone will probably be prepared to buy these. Still, this needs to be negotiated and approved. The rightful recipient of any payments needs to be established. The answer will depend on the set-up of the particular project.

Many items have to be disposed of, but other things must most definitely be retained. Contracts, files and correspondence will need to be kept for legal reasons, and could be needed at any time in the following twenty years. Clearly, the paperwork therefore needs to go somewhere safe. Much information is now stored electronically, but paper files, contracts and written correspondence will need to be archived in such a manner that it can be retrieved fairly easily if disputes arise in the future.

In summary, the period of final completion, an adrenalin-fuelled period of organised chaos, is both highly pressured and highly important for all concerned. The logistician will be tested to the limit, but has an opportunity to make a huge contribution to the successful outcome of the project. As we should never forget, at the end of all the challenges, crossed wires and cross words, the client should take delivery of a building which everyone, from the principal contractor to the smallest subcontractor, can take pride in. This is the goal which gives the logistician, along with all the other professionals and tradespeople who work in the industry, the satisfaction which makes it all worthwhile.

References

Chapman CB, Ward SC and Harwood IA (2006) Minimising the effects of dysfunctional corporate culture in estimation and evaluation processes: A constructively simple approach. *International Journal of Project Management* **24**(2): 106–15.

Egan J (1998) *Rethinking Construction*. The Report of the Construction Task Force. HMSO, London.

Emmitt S and Gorse C (2003) *Construction Communication*. Wiley-Blackwell, Oxford.

Fewings P (2008) *Ethics for the Built Environment*. Taylor & Francis, Oxford.

Harris F and McCaffer R (2001) *Modern Construction Management*, 5th edn. Wiley-Blackell, Oxford.

London K (2007) *Construction Supply Chain Economics*. Routledge, London.

Murray M and Langford D (eds) (2003) *Construction Reports 1944–98*. Wiley-Blackwell, Oxford.

Nowak P, Steiner M and Wiegel U (2009) Waste management challenges for the construction industry. *Construction Information Quarterly* **11**(1): 8.

Ramsey V (2007) *Construction Law Handbook*. Thomas Telford, London.

Chapter 4
Mobilisation and Resourcing the Team

As discussed in Chapter 3, planning is crucial to a successful logistics operation. This is the stage at which outline plans must be drawn up, to ensure that there are no major oversights and to gain a broad view of what will be needed to support the execution of the project. This chapter deals with the next stages. First, there is a discussion of mobilisation. Then there is an overview of approaches to resourcing the project logistics function with the most important assets of all: the people who will deliver the plan.

Mobilisation and site set-up

It would be natural to think that mobilisation is a part of planning, but really it's slightly separate. It does, of course, have to be planned for, but it is essentially a separate discipline and it has its own rules.

During mobilisation, you will probably find yourself located away from the site, either as part of the main contractor's head office, or co-located with a client, or in a short-term office let. Nearer to the start day, the team will move to site. At this stage, some of the logistical tasks described elsewhere in this section, such as setting up temporary accommodation and security, will need to be done.

However, the logistician will probably not be in a position to procure all parts of the logistics package when the project first reaches site. Due to the nature of the common forms of contract, it is very likely that the principal contractor will be in either a two-stage contract with the client or a temporary construction management role with a provisional construction management agreement (CMA). The constructors may not be fully signed up because of either politics or perhaps uncertainty in the development market. The client may be waiting for planning permission, design approval, finance or for a tenancy agreement. Many things can still be unresolved at this point.

There will be much talk of what the team is going to do, and the team will be steadily growing. Still, at this stage, nothing can be set in stone. The project is set to amber, and will be, to a great extent, waiting in limbo. This means that the logistician – along with other members of the team – will have only very limited authority to make purchases. At this stage, the whole project can stall or even stop for a range of reasons. This tends to be a particular problem during recessions. Projects are talked about, everyone is lined up and tenders are prepared. Any necessary demolition or soft strip may even take place. The project, at this point, can quite literally be nothing but a hole in the ground, encased by a hoarding from the demolition contractor that the logistician will need to either take over or replace.

Since there is no green light to spend or commit resources, the logistics manager may well have to buy in early services on a limited spend. Some contracts may be on a weekly basis with an ability to terminate with 24 hours' notice. Once a project is properly underway, it's very unusual for it to be put on stop, but at this point everything hangs in the balance.

It can be an exciting but nervous time. There is the initial enthusiasm whilst everything is being planned, but this is taking place against a background of total flux. There is so much to do, but no authorisation to create the spend which would get things really moving. The client's quantity surveyor will be particularly busy at this stage, ensuring that their employer does not get unnecessarily committed before they finally press go.

Nonetheless, despite the reluctance to spend, temporary accommodation will have to be procured, even if it not the finished article. The project will also need some very basic pieces of kit, such as fire extinguishers, personal protective equipment (PPE), toilets, toiletries, coffee and tea. Clearly, the team will also need desks and computers, which will probably sit side-by-side with crates taken out of storage from a previous project. An office also needs photocopiers, printers, toner and paper, alongside other items of stationery. The office will need phones. Whilst the spend has to be minimal, the amount of items needed for even a skeleton staffing level start to mount up. The office manager, if there is one, will organise many of the purchases, but the logistician will be responsible for getting the items to the point of use. Since the budget is so constrained, a certain amount of scavenging will often be undertaken by various members of the team. Calls are placed to former colleagues, particularly any that are deemed to owe a favour, and particularly those with a site shutting down. This exercise can yield anything from fire extinguishers to toilet roll. Some of the limited budget may well be spent on a 'man with a van' who can collect whatever happens to be available.

The logistics manager, regardless of whether they are the construction manager, junior manager or logistics contractor, is likely to find this phase reasonably stressful. The challenge is to provide logistics on a shoestring whilst maintaining the company's professionalism. It's also vital to use this time to get to know the neighbours, particularly the neighbouring businesses and schools, hospitals etc. Otherwise, you will usually meet them for the first time when you block the road off. This is never the best way to get onto a friendly footing. When the situation is in limbo, the uncertainty can seem to create an infinite amount of time. Unless there is a stop, however, the logistician will be fully occupied all too soon. Contractors and materials will be flooding onto site, and the project will be in full swing.

Resourcing the logistics team

As we have previously stated, the very word 'logistics' can create confusion, perhaps because the function has such a potentially wide remit. At one end of the scale, it can involve high-level supply chain management, but its activities extend right across the spectrum, from strategic planning to emptying waste bins.

The main focus of the remainder of this chapter, however, is the operational aspects of site logistics management, and options for their resourcing.

The background

Historically, the major construction companies employed a significant amount of direct labour. By the 1980s, however, the trend had moved to much smaller principal contractors, managing a pyramid of subcontractors. This was a macro-scale reflection of an existing shift from hourly remuneration to piecework, an issue that has been hotly debated in construction for centuries.

Employers had long been keen to have the flexibility to reward productivity rather than attendance, whilst the unions traditionally resisted such moves. As long ago as 1850, it was argued by the general secretary of the Stonemason's Union that piecework would induce workers to 'overstretch every sinew to make as much as they can', creating an unsustainable norm of inflated wages which the employers could then swiftly reduce (Wood 1979).

The matter was finally settled in an agreement of 1947, which ushered in 'payment by results' as a permanent option following its deployment during wartime (Great Britain Ministry of Public Building and Works 1947). This set the conditions for a growth in self-employment, which would later be accelerated by the introduction of tax exemption certificates in 1971. Like all changes, this fundamental shift in the system of building brought both benefits and drawbacks, not all of which would have been foreseen by the advocates of the payment by results system, who had merely wanted to be allowed to incentivise efficient production. Indeed, many productive trade contractors are as interested in work–life balance as they are in inflated wages, and will leave site early once they feel they have achieved their target earnings for the day.

The large building contractors would once have directly employed the labour they needed to perform logistics functions on site. With the advent of lean management organisations, whose primary function is often to allocate and manage packages of work, that option became impractical. At the time of writing, a major contractor would be most unlikely to turn to their own human resources department when looking for logisticians or logistics operatives. Direct employment creates liabilities that can be incongruent with the shifting project commitments of the principal contractor, so the natural trend was towards identifying other options. As we shall see, however, the current pyramid of cultural relationships, which has created what is often described as a fragmented industry, can also foster a fragmented and inefficient approach to logistics management.

Contractual norms and logistics management

Naturally, whilst there is not an embedded tradition in construction of professional logistics delivery, the practical provisions for performing logistics functions have always had to be created. Materials have to be delivered, and waste has to be removed. Numerous support functions still need to be fulfilled.

In construction, many of these functions are grouped together under the contractual banner known as 'prelims'. This makes the principal contractor ultimately responsible for the provision of umbrella services such as security and temporary accommodation. However, due to the nature of construction contracts and the allocation of risk, many logistics services, such as housekeeping, will then be parcelled up in the trade packages let to subcontractors. In practice, this means that scores of operatives can be at least notionally involved in logistics functions at any one time, often without any coordination.

As well as creating substantial inefficiencies, this practice also tends to foster a 'make do and mend' approach to logistics, rather than the carefully planned, professional function argued for in this text. The detrimental effects of this have been widely recognised by analysts, but as yet the alternatives are not widely adopted.

As Ashworth (2005) argues, 'The industry lacks an integrated approach to its work.' He goes on to cite the industry's proliferation of micro-firms and adversarial culture as the cause of problems such as inefficient use of labour, high wastage of site materials and 'functional inefficiencies' which cause high running and maintenance costs, as well as low profitability and a poor public image.

Clearly, at least some of these issues could be tackled best by improved logistics management. However, an overarching logistics service will, clearly, require a logistics budget. If you've created a budget, you've created an opportunity to go out and procure professional logistics. But while that sounds reasonable in principle, the form of contract can make that very difficult in practice.

Typically, there will be no dedicated budget line to deliver logistics within the cost planning for the job. There may well not be buy-in from the client's quantity surveyor. Ideally, the author would argue, companies should plan and agree a logistics budget at a very early stage. In reality, even if there are specific allowances for logistics, it is likely to be one of the first items to fall by the wayside if there is any downward pressure on the overall budget.

In theory, it should be possible to identify the logistics budget by asking the main contractor to identify what costs they have allowed for the relevant preliminaries. In practice, no detailed costings may have been taken – it's all part of the traditional art of construction pricing, which can provide ballpark costs as an umbrella for a number of variables. Further, since these costs are then passed down, you may find that a number of trade and subcontracted firms have notionally allocated 0.25 of an operative's working day to the tasks – making it incredibly difficult to ring-fence any funding.

Due to the complexities of achieving change, it remains normal to pass the risk on to the individual trade contractors. This creates a self-perpetuating but, it is argued, unsatisfactory situation. Construction logistics, therefore, has yet to become a fully fledged professional function, and its value remains widely unrecognised. Things are done, but they are often done inefficiently and expensively.

Issues in current practice

Besides, logistics is not generally seen as the core business of the contractor. They have incredible skills in creating buildings, but often little interest in the routine process management that is at the heart of operational logistics. Therefore, logistics remains a kind of 'grudge purchase', and is rarely accorded systematic resourcing.

The current approach, which commonly includes practices such as making contractors responsible for such functions as removal of their own waste and delivery of their own materials, can be an extreme example of contractual devolution of risk and responsibility, and it can cause correspondingly extreme challenges of coordination.

It does not take extensive site experience to anticipate that this practice can frequently lead to inconsistencies and possibly even gaps in the provision of important services. If a particular contractor is responsible for maintaining a handrail, who is responsible for it after they've left the site? Most contractors are not involved for the whole duration of a project. Therefore, certain tasks will clearly have to be given to others as they complete their package of work.

Furthermore, this system is poor at making provision for logistical issues where responsibility is contested. To take a common example, if rainwater falls through a hole in the roof, is it to be considered part of the remit of the trade contractor who left the hole or is it a problem for the firm who laid the (now badly damaged) flooring?

Put in these terms, the value of a team which can give a holistic response to such emergencies becomes obvious. Currently, the construction manager has the option of offering a trade contractor a variation for such remedial tasks, but there is no guarantee that the trade contractor will want the work, and it is not uncommon to see such offers refused, particularly if the task is one that their employees are likely to perceive as menial. Why should the highly skilled individuals they employ necessarily wish to perform semi-skilled tasks?

That said, these same skilled operatives are often left responsible for picking up their own rubbish, which is an uneconomical use of their time. Considering that they might be attracting a charge rate of perhaps £25 per hour, this is not a financially prudent use of resources. The specific problems with this approach to rubbish collection are further discussed in Chapter 10, but the frequent outcome – that rubbish gets left, and there are disputes about who created it – is one of the triggers for another frequently deployed strategy for waste management.

To avoid unproductive conflict arising from situations such as unclaimed rubbish, the project manager may elect to buy in agency staff to perform routine tasks such as rubbish collection. After all, the project manager or a deputy will be de facto in charge of logistics on a traditional site, and therefore will not usually be able to devote the necessary time and attention to logistical issues.

Buying in agency labour is a quick fix, which means that routine tasks are dealt with. Unfortunately, this route often involves buying in labour from the most unreliable, casualised and demoralised segment of the entire labour market. The nature of the contractual arrangement means that there is no mechanism for developing the skills and loyalty of the workers.

By buying in such help, the project often ends up down a route of spending an uncapped amount of money in order to deal with the disadvantages of fragmented responsibility for the logistics function. This will be done reactively, in response to problems and to diffuse potential disputes. The strategic planning underpinning core logistical tasks may well, therefore, be minimal.

The use of unskilled staff, with no specialist logistician to lead them, is unlikely to create an environment of optimal productivity or problem-solving. The agency workers, in addition to lacking any sense of organisational identity or opportunities to progress, will almost certainly not be incentivised to achieve targets, and may be concerned that they will be laid off if they work too quickly. This means that the response to any ongoing problems may well be to increase the number of agency staff on the site, rather than establishing sound leadership to maximise the prudent resourcing of the many logistical tasks a site requires.

This situation can lead to such inefficiencies as skips leaving site half-full – the project is literally paying for the removal of fresh air if they are not filled. Nevertheless, it's quite common for busy managers to order their removal purely because they're in the way.

As well as creating inefficiencies in the removal of waste, the situation will often also constitute a missed opportunity to put a check on the over-ordering of materials. If a contractor overestimates the amount of material required, or if the material is damaged through bad handling, the cost of the entire order, plus its standard mark-up, will still ultimately be charged to the client. It is common for unused materials worth hundreds of thousands of pounds to be consigned to waste, unused, by the end of a major project. For that reason alone, there is a compelling commercial case for change, but this is only one aspect of professional logistics support. It also has the potential to create many other synergies and savings as well.

The specialist construction logistics contractor

The passing of responsibility on to trade contractors, the employment of agency staff or a mixture of the two are common attempts to solve the issues of logistics. The other option is to hire a specialist logistics contractor. For the reasons outlined above, it's still a relatively uncommon option, but the commercial benefits are potentially significant. After all, every other major industry recognises the benefits of professional logistics management, and it's surely only a matter of time before it becomes normal practice in construction.

Like any business function, logistics carries an element of uncertainty and risk. Therefore, there's a strong argument for devolving that to a specialist contractor. Currently, many clients and their quantity surveyors have a perception that this will cost more than passing the risk down to the trades, although in actuality that route is far less transparent. The client pays for huge inefficiency, and possibly 20% more material than required.

If the client allows for a specialist logistics tender, by contrast, there is an opportunity to define the scope of works. The tender can specify exactly what is required, supported by a cost schedule. The client can also then specify what they want in the quality of the personnel provided, in terms of their skills and qualifications. This route also helps to ensure that all necessary method statements and safety, quality and risk assessments are in place. A good logistics contractor will also predominantly be using their own workforce, and therefore morale, loyalty, skill levels and teamwork are likely to increase.

Direct employees are clearly a desirable asset, but the use of a specialist contractor can, if deployed correctly, achieve substantial cost savings.

Take a standard £50 million commercial development with, say, a 100-week programme. There might be 30 trade contractors on site at any time, each with two labourers. If these 60 labourers can be replaced with a dedicated logistics team of 20, then substantial savings start to accumulate. By systemising the delivery of materials, productivity can be increased by perhaps an hour per trade per working day, which could save 150 hours a week. This is a significant enough saving to accelerate overall completion time by perhaps five weeks. In fact, it's not unreasonable to expect that these can be 2–4% of the overall cost of the project. Obviously, this is an indicative calculation, but six-figure, or even sometimes seven-figure, savings have been achieved on real projects.

A specialist logistics contractor is likely to have committed considerable resources to multi-skilling the team and will have a number of highly qualified staff. Their ethos will be to get the maximum benefit out of the working day, and so it makes sense to have flexible staff with a broad range of competencies. In addition to increased control over the quality of the workforce, it will give the client or principal contractor a senior point of contact for all issues relating to logistics management and the benefits that can derive from specialist leadership of the logistics team.

It is a far more effective way of dealing with site housekeeping, and those health and safety issues that result from poor housekeeping, and other issues which are common to the whole site rather than to any of the subcontractors. The logistics team can come together to form an effective emergency response team, and be assigned specialist duties to perform if there is a major incident.

Indeed, their contribution can be so useful that one of the issues dedicated logistics teams have to deal with is the way they can end up doing far more than they were contracted to do. The logistics manager has to be extremely wary of scope creep. Due to the nature of the work, the scope of tasks required can slowly but surely expand way beyond the terms of what was originally agreed. Whatever the problem, there is a temptation for the construction manager to ask a logistics operative to sort it out. If this happens six or seven times in a day, core jobs can get left. Therefore, the team needs to be disciplined, although not too rigidly so – it helps to have a 'can do' approach. Still, there is such a thing as being too helpful, and that can affect the commercial viability of the project. If the issue gets really strained, the manager may have to instruct staff to record exactly what they have done that day. Nevertheless, a certain allowance for unplanned work can avoid this in most cases.

After all, many of the more experienced construction professionals have regrets about the passing of the comprehensive in-house team of a generation ago. They have seen the potential pitfalls of using casual labour, and realise that the logistical problems created by the demise of the traditional large contractor have never been comprehensively resolved.

It's the author's belief that logistics contractors can, at their best, create what construction companies would have had in-house 40 years ago. Logistics managers can perform functions similar to those of a traditional foreman. On smaller sites, many supervisors and gangers of today are undoubtedly able logisticians (whether or not they are aware of it), but many larger sites would be likely to benefit from specialist assistance in managing the complexities of modern-day logistics.

Whoever is running the logistics function, whether they are a specialist contractor or the deputy project manager, they should never lose sight of the importance of their team and its contribution to the site. There is always a danger that the logistics team will be undervalued, perhaps even to the point of being treated discourteously by other site workers. After all, we do not accord much cachet to the removal of rubbish or cleaning toilets.

Nonetheless, they are as vital to a well-run construction site as the grounds staff is to the FA Cup Final – the unsung heroes of the project. This is a key message to communicate to the team. They should be recognised for their contributions, given quality working gear and respected for their skills and dedication, particularly when performing some of the most unpleasant tasks on site. (The author always negotiates a premium for toilet cleaners. It's routinely perceived as a minimum-wage task, but the reader might usefully reflect on what hourly rate they would find attractive enough before they would shoulder such a duty on a construction site.)

By constantly reminding the logistics team that they are as much a part of creating the building as anyone else, by fostering teamwork and multi-skilling to ensure job-rotation, a 'menial' job can be transformed into a rewarding career. Training and a professional approach to the job gives members of the team transferable skills and the status which, after all, we're all entitled to in return for a fair day's work. Figure 4.1 shows

Figure 4.1 The logistics team at the £150m Manchester Arndale extension project

the logistics team assembled for the £150 million extension project at Manchester Arndale.

Employment and training

One important consideration when staffing the team is the significant role the construction industry can play in regeneration. Not only do the projects themselves have an enormous role to play in reshaping local communities by providing infrastructure and buildings but also construction firms can directly assist in the regeneration of the community, whether or not it's a regeneration project. On most sites, firms can add significantly to the economy of the local area by recruiting and sourcing locally, by using local suppliers and by encouraging contractors to recruit both skilled and unskilled labour from the local labour pool.

On the bigger projects, the constructors will have local regeneration agencies seeking involvement from the outset, and planning may be contingent on certain local employment and training targets. Although the employment of local labour cannot be a direct requirement of planning permission, it can be made part of an agreement under section 106 of the Town and Country Planning Act 1990 (Macfarlane 2000). In any case, it is good practice to encourage the community to get involved with a project, and to try to ensure that there are job opportunities for local people.

That said, it may be quite difficult for the major trades to take on local staff. There is usually not enough time to train skilled people, since it takes months or years, unless it's a really large and long project. This is an opportunity for the project to win friends in the community by creating jobs through the logistics contractor.

The logistics team is on site from almost the beginning and will be one of the last teams to leave. In contrast, many of the trades will only be involved for a few months, for a particular phase of the project. Furthermore, the semi-skilled nature of the work done by the logistics team means that they can be trained quite quickly. People from outside of construction have many untapped skills from previous jobs, or hobbies such as DIY, and can be trained for a variety of skills reasonably quickly.

Therefore, it is possible to take a raw recruit with aptitude (hand-to-eye coordination and other attributes that can be easily tested) and train them to be a member of the logistics team in 4–6 weeks. Within that time, a new member of the team could be trained in hoist driving, materials handling, waste recognition, fire marshalling and first aid plus perhaps another one-day course on general health and safety awareness. Very quickly, you have given local trainees a range of skills and employment opportunities. Figure 4.2 shows Hiab training being undertaken off-site.

Figure 4.2 Off-site Hiab training

When the recruits are trained, the author has then found it helpful to use an informal mentoring system as they begin on site. (Many operatives would baulk if formally assigned a mentoring role, but will in fact do the job superbly if merely asked to look after a new colleague for a week or so.) The logistics manager will want to use the more experienced staff in management or supervisory roles, who can then assist with the ongoing training of the new recruits. The other advantage with logistics is that the logistics firm may sometimes acquire a role, such as provision of security, in the completed building, in which case the project may ultimately provide permanent work for some local people.

Local staff will often have more investment in the project than average. They have a vested interest in good outcomes, since the project is an addition to an area that they and their families may well be very emotionally attached to. It will usually be easy, therefore, to get them to take pride in their work, and to get them to buy into the need for those acting as a first point of contact on site to provide excellent customer service to visitors. Their wages will also put more money into the local economy.

This does nothing but good for the project. If you take people from the local community and make them a part of a project, they will often derive enormous pride and satisfaction from their involvement. The local knowledge they hold often brings many advantages in sourcing and securing local services and making influential contacts.

References

Ashworth A (2005) *Contractual Procedures in the Construction Industry*, 5th edn. Pearson Education, Harlow.

Great Britain Ministry of Public Building and Works (1947) *Payment by Results in Building and Civil Engineering During the War: A report on the operation of the payment by results scheme applied under the Essential Work (Building and Civil Engineering) Order, 1941, during the period July 1941 to March 1947*. HMSO, London.

Macfarlane R (2000) *Using Local Labour in Construction: A good practice resource book*. The Policy Press, Bristol.

Wood L (1979) *A Union to Build: The story of UCATT*. Union of Construction, Allied Trades and Technicians, London.

Chapter 5
Materials Delivery and Handling

In 2009, the Construction Products Association (CPA) stated that the product (materials) sector of the construction industry had a turnover of more than £40 billion, constituting 40% of total construction output (Construction Products Association 2009).

This figure of 40% is also roughly the percentage of a project's budget which will be spent on materials. Therefore, this is one of the most important areas within the logistician's remit. For as long as a project is active, materials management is likely to absorb a significant part of each working day: the challenges created by this function are constant.

Materials management is a multi-faceted activity. Not only do materials have to be brought to site and be unloaded, they also have to be moved around the site to arrive in the right place, at the right time, so that the project can progress. Materials management involves almost everyone on site; staff at every level will be involved, either physically, administratively or managerially, to deal with this crucial flow of resources to the site.

Contractual issues and current practice

Due to norms in contractual relationships over the last 15 years or so, the unloading and movement of materials (both to storage areas and ultimately to the workface) has largely rested with the trade contractor who procured them. Either they or their subcontractors have to take responsibility for protecting their materials. Occasionally, but not often, the trade contractor will have a materials handling team, who will be responsible for unloading and moving deliveries. More usually, it will be done by either the tradespeople themselves or by unskilled agency labour.

This may seem reasonable at first sight. In practice, however, the complex and fragmented nature of the supply chain renders materials deliveries chaotic and inefficient on a large proportion of construction projects.

Let us pause a moment to consider the nature of the problem. On a typical city centre office block, there will probably be 30 or more trade contractors. Potentially, therefore, you might have 150 subcontractors. Many of these subcontractors may have sublet their portion of the contract down a further tier of the supply chain. Therefore, you have upwards of 200 firms on a site. They will be dealing with upwards of 200 suppliers.

The products they are purchasing will depend on what has been specified by the architect. The top tiers of the supply chain, the principal contractor and main package trade contractors, might buy some materials directly from a supplier, or they might procure via the lower tiers of the supply chain.

When we consider that there will be upwards of 200 suppliers, the manufacturers and wholesalers, further still down the supply chain, we start to see how complex – how chaotic, even – the logistics of materials delivery can become. All parties are apt to use multiple suppliers. If you're the cladding contractor, for instance, you might buy your steel frames from Supplier A, your glass cladding panels from Supplier B, your gaskets from Supplier C, your brackets from Supplier D. You might well also be paying Supplier E to put components together off-site, and Supplier F to manage the delivery to site once the items are assembled. When we think about the number of firms on site, we begin to get an idea about the sheer complexity of all the contractual relationships on a project.

Due to this complexity, it is almost impossible for any one party to have a comprehensive knowledge of who should be doing what at any given moment in time. The principal contractor will be dealing almost exclusively with the tier-two package contractors, and many site managers, particularly those who are younger and less experienced, will have very little idea about the identities of firms further down the supply chain. (Right at the bottom of the contracting pyramid, there may be many self-employed individuals or small groups, and no one on a project will be familiar with all of them.) Indeed, principal contractors have no direct contractual relationship with these lower-tier subcontractors – a fact which can cause serious problems on a project if a tier-two or -three contractor becomes insolvent.

This multi-layered system breeds many inefficiencies. Aside from the sheer confusion, it means that many parties in a chain will be selling services and materials cost-plus, which can result in five or more layers of cost being added to materials transactions. It's both messy and expensive. All that said, the contractual pyramid looks unlikely to end in the foreseeable future. The logistician, therefore, needs to be aware of the problems inherent in the system in order to manage them to the best of their ability.

So, how is this to be done? First, the logistician needs to map the relationships as accurately as possible. Then, they need to try to manage the communications and practical dimensions of working out how to get the materials to the people that need them. They need to be aware that communications which emanate from the top tier can get altered and misconstrued as they're passed down the lines, and they must try to keep in touch with what's going on at the other end of the pyramid. As previously mentioned in Chapter 3, 'management by walking about', which gives the logisticians the chance to talk to a wide range of people, is crucial to logistics, because that's the best way of finding out what's actually happening – and what's actually needed – on the ground.

Getting materials to the site

Most construction sites, particularly those in city centres, have to operate under significant constraints in terms of their delivery capacity. There will only be a limited number of bays that can accommodate lorries, and a limited number of hoists and staff members to unload them. Crucially, there are only a limited number of hours in the day when the deliveries can take place.

Due to these constraints, the daily site meeting, which will typically take place in the late afternoon, will often resemble a City trading floor. Robust negotiations and trade-offs will be taking place, as all the contractors try to ensure that they will get the delivery slots they need for the next day. This will not always be easy. Some contractors will have booked their slots a week in advance, and they will not be keen to surrender them. Everyone is acutely aware of the pressure on the system and so will often book more than they need, to create a buffer for their own deliveries. This is an understandable strategy for the individual contractor, but it's not at all good for the progress of the overall project, since it puts even more pressure on an already strained resource.

Because of the scarcity of slots, a driver arriving late for any reason can be turned away. Therefore, the temptation is to book a two-hour slot for a delivery which might only take 45 minutes to unload, in order to build in a buffer. If the trade contractor's staff are to be the ones unloading the lorry, they tend to wait until the last possible moment before they leave the workface to unload it. They are likely to be tradespeople, and they will not want to be hanging around the unloading bay in anticipation of the arrival of the lorry. This means that the driver will be the one kept waiting – which puts more pressure on the limited time and space in the bays. The difficulty with securing slots also underpins the temptation to over-order materials and create a buffer. This means that unneeded extra materials, ordered just in case, must somehow be stored on an already space-constrained site. This leads to double handling (and poly-handling), which increases the likelihood of damaged materials, all in the name of programme certainty.

The primary logistics function is to have materials arrive when they're needed. The logistics manager will usually have 10 hours per day where unloading is permissible. Given that, the ideal is to have the first unloading at 08:01, and the last at 17:59. If the schedule is not respected, it upsets the neighbours and the site can lose its permits to unload. Sometimes, mutually beneficial arrangements can be reached. A hotel will not like deliveries next door during prime check-in times, but might be happy to allow evening deliveries, when the guests are all at dinner. As with everything, the logistician needs to work with the site neighbours to understand the constraints of their particular businesses.

On the average site, the precious unloading hours are not utilised in an optimal manner. The logistician will make every effort to smooth delivery patterns, but current traditions have to be challenged. Hauliers usually want to dispatch deliveries early, their aim being to get in first and beat the rush. The upshot of this strategy is, generally, early-morning chaos. In the afternoon, by contrast, bays will be standing empty. What you try to do with logistics is to create a regular heartbeat, with regular and consistent arrivals. This process is called 'smoothing'. Not long ago, this all had to be arranged using pen, paper and a whiteboard. Now we are blessed with software for scheduling, where contractors can book resources for unloading as well. The best systems will tell you where there is a potential conflict, for example where the hoist has been booked out already.

When planning deliveries, the first thing the logistician needs to understand is how many unloading bays are available, and what facilities for unloading need to be provided. There tends to be little space for such operations, which will be relying on a crane or forklift. The options are somewhat limited by the very nature of construction. You are usually limited to unloading from the kerbside of the public highway, although in special circumstances the Local Authority may be persuaded to allow a lane closure to create an unloading zone or pit lane. In a city centre, you will rarely have the luxury of bringing trucks into site and off the public highway. This, of course, is not the case for housing developments or greenfield sites (or, indeed, consolidation centres (CCs), as Figure 5.1 demonstrates with a range of delivery vehicles parked up at the Colnbrook Logistics Centre at Heathrow Airport).

Your options to lift materials off the truck will be by either forklift, crane, beam winch or, as a last resort, muscle power. If you are unable to build an unloading platform at truck height, which would allow you to unload directly into a hoist, your options are crane to access platform (Figure 5.2), beam winch or gantry crane (Figure 5.3) to a level one gantry, or forklift to ground-floor hoist.

Figure 5.1 Delivery vehicles parked up at the Colnbrook Logistics Centre at Heathrow Airport

Figure 5.2 Logistics operative at Skanska Barts project moves materials using a motorised pallet truck via a goods hoist

A gantry crane will have no more than a 1.5 tonne lifting capacity. It goes out over the roadway at first-floor level, bringing loads into the site footprint onto a first-floor loading platform. This can avoid unloading onto the street, which would otherwise render the project guilty of obstructing Her Majesty's highway. Using tower cranes, or sometimes mobile cranes, will allow you to take materials directly from the back of a truck. Materials can then be landed on the roof, or into floors via an access platform. There are a variety of off-the-shelf access platforms available from plant hire companies. Alternatively, you can build one using scaffolding. Access platforms are versatile and easy to move, but only provide you with an option whilst craneage is available. Unloading from the street with a forklift (or by hand) is problematic, as you may cause obstructions to other road users and pedestrians. Some trucks now come with a forklift already on board, or have on-board cranes (known as 'Hiabs').

Whatever the unloading system and the methods for scheduling its use, it must be remembered that, however sophisticated the booking system, if the truck doesn't arrive, then all the contractor has done is scheduled a missed opportunity. This is at the root of a huge amount of conflict. If a driver is half an hour late, the delivery is likely to miss its slot. The truck driver may very well have been delayed by problems on the motorway network or such unavoidable occurrences as a puncture, and feel that they have been disadvantaged through no fault of their own. Still, since the construction manager will be acutely aware of the need to support scheduling discipline, the bay supervisor may be told to send the delivery away. This seems harsh, if there was a genuine reason for the hold-up. On the other hand, if other deliveries are bounced off the schedule to fit it in, there will be no discipline at all in deliveries. In many aspects of logistics, you have to guard against process decay. If that happens, no one will follow the rules. The logistician should be aware that the same construction manager who sends deliveries away without exception may soon be the one asking for exceptions to be made in response to pleas from the package contractors. Under pressure, it's easy to lose discipline. This will certainly militate against an effective delivery strategy.

That said, it's hard not to have sympathy with drivers who have just been told they can't unload. Drivers delivering ready-mix (see the discussion of concrete below) are particularly likely to look askance at any attempt to impede their progress. On the average project, in the author's experience, logistics staff will ban perhaps a dozen drivers for verbal abuse or worse. Usually, there is no malice, however; it's just a pressured and conflicting environment where everyone wants to protect their own interests.

The delivery bay is a robust environment, and the person in charge ideally needs both thick skin and the verbal skills of a world-class diplomatist. If the fault is not with the driver, a waiting truck can charge demurrage costs. More compelling even than that can be the extra time constraints created by the driver's legal working hours, which are monitored by the vehicle's tachograph. One driver at a major London site, on being prevented from unloading, parked up to create a blockage and waited for a police escort to move him, on the grounds that his taco had run out. Contractors and construction managers alike are regularly pulled out of meetings to deal with such tense situations. A good loading bay supervisor will work miracles, although this may involve having waiting lorries round the corner with their engines running. This can prevent major delays and confrontations, but it's still far from ideal. It does nothing for construction's image, or its environmental performance, and can bring the project into disrepute.

Due to the complexity and constraints of the current system, getting materials to the right place at the right time in the right condition is achieved perhaps 30% of the time on average. A figure of 40% is currently state of the art (Transport for London 2007). If 20 trucks are booked in, six will arrive on time, seven will be early or late and seven will not arrive at all. Two trucks that weren't booked in will turn up whilst you are waiting. This is representative of a typical day.

It is the author's belief that CCs, discussed later in this chapter and, at more length, in Chapter 11, represent the future of materials management in construction. As yet, they are in their infancy, but they constitute an approach that can remove much of the pain and inefficiency from this key logistical challenge in construction. As it is, an inefficient system is staffed by reluctant tradespeople or untrained agency staff, and there are strong drivers for each individual company to overload a struggling system in order to ensure that they meet their own obligations. The multitude of stakeholders involved from truck to warehouse also tends to ensure that few have the right equipment to transport their delivery. Culturally, construction has an ingrained tendency to rely on muscle power.

The challenges of materials management apply equally to larger and smaller sites. The only difference is that of scale. Due to the specialisms involved in almost any construction project, there will normally be about 30 main packages of work, regardless of whether the overall project is worth £10 million or £200 million. Therefore, the same principles apply to managing the logistics.

Common user plant: tower cranes

Common user plant, as its name suggests, is provided for the use of everyone on a project. This is a key part of the logistician's remit. Once upon a time, before the dawn of modern logistics, this used to be confined to hoists, temporary accommodation, access points and cranes. As construction logistics has grown more sophisticated, however, the range of common user plant has grown. Some of the aforementioned items are discussed elsewhere, but common user plant often refers to equipment used to move and distribute materials (as well as equipment and people), so this seems to be the logical place to discuss it.

One of the most important items on the growing list of items sourced as common user plant remains the tower crane. This is not the place to discuss the deployment of tower cranes in detail, because specific information is available for this specialism. However, the logistics manager will be very much involved in working out where any cranes are best placed, and may be involved in the logistics of the actual erection of the crane. Having said that, the cranes will probably be procured by a specialist crane contractor or the frame contractor. This element of the project will also need extensive engineering input. Such technical issues are best dealt with by a suitably qualified person, and this is not likely to be the logistician. Nevertheless, the logistics manager needs to understand the basics.

There are two main types of tower cranes, the traditional fixed jib and the luffing jib.

The essential points to establish for tower crane operations will be the weights required to be lifted, the height of the building (since the tower crane will clearly need to be higher!), the reach of the tower crane and the place where it will pick up and put down loads. The loading and unloading will often cross buildings and streets. During the site layout planning stage, working out where to put the tower cranes will feature very strongly. The cranes arrive early in a project, and a building may, quite literally, be constructed around it. Therefore, it can sometimes be quite difficult to dismantle and extract them! Getting the crane *in situ* is a major operation – a crane has not only mast sections and a boom but also counterweights. A crane is, in itself, a significant construction – which is why you need a crane to erect a crane.

Therefore, the best location for a crane depends on a number of factors. It will need a suitable base, clearly, but the location will also depend on where the reach of the crane might over-sail. A fixed jib crane might over-sail beyond the footprint of the building, perhaps over roads and over other buildings. You can't over-sail either without permission from, respectively, the Local Authority and the building owners. These agreements are known as 'over-sailing rights', and are subject to negotiation, with the agreement usually involving the payment of a fee. A construction project has no fixed rights to over-sail, and the owners of buildings underneath its reach have an absolute right to refuse permission. Because of this, the costs involved in agreeing these rights can be significant. Since the crane can't be erected until this is settled, it's important to negotiate these rights as early as possible.

Where over-sailing rights are withheld, one alternative is to use a luffing jib crane, but these have a lower capacity in terms of weight and load reach compared to fixed jib tower cranes. Another way of overcoming such a situation is to be imaginative about where to put the crane. The author remembers one job in the City of London where such an issue arose. An ingenious solution was worked out by the frame contractor and engineers. They worked out that they could build a concrete core using jump form. They created a specially strengthened core, so that the crane could sit on top of it in the middle of the building, albeit without its usual mast. As the saying goes, necessity is the mother of invention. In construction, the constraints of a project will often be the driver of extremely creative innovations.

However, the negotiation of over-sailing rights will still generally be the preferred option. The crane should ideally be positioned so that it can pick up from all loading points and put down in as many locations as possible around the site.

Another important issue with tower cranes is, of course, that the logistician needs to make sure that the communications are thoroughly tested. The radio frequencies must be able to pass information between the crane driver in his cab and the banksmen and slingers on the ground.

The whole issue of tower crane operations is strictly governed by the Lifting Operations and Lifting Equipment Regulations 1998 (LOLER), which apply to all lifting and lowering operations (Holt 2001). There will need to be an appointed person under these regulations, who will have to be a member of the principal contractor's team. When there is more than one crane, the project will also need a tower crane coordinator. The coordinator should be from the same organisation as the one providing the drivers and banksmen, but will report directly to the construction manager. A tower crane will be erected in sections, and each section needs to be checked and approved by the project's health and safety specialists.

It should be noted that at the time of writing there have been some high-profile incidents involving tower cranes collapsing. It would not be appropriate for the author to comment on specific incidents themselves, but all accidents remind us that safety has to be right at the very top of the agenda. All plans involving tower cranes need to be double-checked to ensure that they are safely procured and erected.

As obvious as it may sound, it's particularly important to ensure that, if there are going to be two or more cranes on a project, they are sited where they are not going to collide. If the cranes are of different heights, one jib may be over-sailing another under certain circumstances. Collisions have happened. When planning the locations of cranes, beyond the necessary engineering considerations which are subject to the laws of physics, the logistician must never forget Murphy's Law. Simply stated: anything with the potential to go wrong, will go wrong. In many areas of construction logistics, and with particular reference to construction safety, this is the great commandment.

The logistician will become very involved during the phase when the tower crane needs to be erected and then operated. During the assembly of the tower crane, its components will be delivered by trucks. Therefore, the first thing the logistician needs to do is to work out whether these should be escorted loads, because of their width or length. Then the type of mobile crane needed to assemble the tower crane will have to be identified. Will this mobile crane fit through the streets, avoiding height- or weight-restricted areas around site? The logistician must also consider the kit needed to assemble the tower crane, arranging any necessary road closures with the Local Authority and the tower crane provider. The logistician will have overall responsibility for checking that all these things are done, as well as checking that the necessary permits and oversailing agreements are in place.

Experienced tower crane providers will be familiar with the procedural requirements, but this is not a reason for the logistician to step back from the arrangements. The tower crane provider might be adept at arranging permits and escorts, but this is not a reason to assume that they will have taken a holistic approach to relationship management with other stakeholders. Therefore, the logistician must ensure that suitable liaison takes place, especially with the site's neighbours. If there are to be road closures, it's important to ensure that this won't affect their own commercial plans. Closures can be arranged for off-peak times, but the components of a substantial crane might need to be unloaded over several weekends. It makes sense, therefore, to be proactive about keeping all those affected informed.

One of the other issues with tower cranes, this time relating to their operation, is that the needs of different contractors can lead to clashes in terms of demand. In particular, once the frame contractor has finished on site, the logistics manager needs to take control of a situation where there will be, almost inevitably, competing demands. The logistician needs to be aware – and realistic – about the dynamics that can be brought to bear.

Common user plant: lifts and hoists

Goods lifts are another key piece of common user plant. Predominantly, rack and pinion hoists have been used, and they can still be seen trundling up and down outside construction sites all over the UK. Now, however, lifts also come in many shapes and sizes, and have many purposes. There are goods-only hoists (Figure 5.2), for instance, and there are multi-purpose goods and passenger hoists.

The logistician must ascertain which type is most appropriate, which will, as with so many facets of construction logistics, depend on many different project-specific characteristics. How many people will be on site, and what will they be doing? What is the anticipated flow of materials? To some extent, the selection and quantity of hoists will be dictated by the experiences and preferences of the construction manager. Still, the logistician should look at the programme and build type, as well as at the labour histogram. Using those figures combined with the experience of the overall construction manager, a reasonable estimate can be made regarding the correct number of hoists.

Figure 5.3 Materials lifted to gantry level at Heathrow airport

This estimate will be assisted by computerised modelling techniques. There are methods of data capture and analysis that, by looking at the programme and its peaks and troughs, can forecast both materials flow and what level of hoist utilisation will be required to service it. Calculations of hoist requirements, whatever the method used, will generally be working on the assumption of a 10-hour working day delineating hoist availability.

Sometimes, this analysis will tell you that you require more hoists than it is physically possible to put on the job. In the author's experience, that may well be telling you something about the programme. Providing feedback of this nature can cause consternation with programmers, but it is a useful indicator of overall feasibility. Programmers are highly skilled at what they do, but their calculations are generally based on assumptions about how long it will take to fit various types of materials, rather than on how long it will take to deliver and unload it. In fact, the constraints relating to how much kit can be physically got to site are extremely important to the programme overall.

It's also important to understand the detail of materials volume when planning hoist requirements. Clearly, it's important to find out the biggest items that will have to be lifted, but it's also useful to know how many of these there are. Bigger hoists generally mean slower and less frequent numbers of lifts.

In an ideal world, a hoist would take materials up once, and come down loaded with waste. In reality, materials will sometimes go up and down in a hoist several times prior to distribution. Hoists can also create inefficiencies when they are people carriers – even if a journey could be completed quicker on foot, waiting for a hoist to come, should one be available, is the most natural thing in the world. There is a danger that a project can end up with a very expensive passenger lift that is doing very little to improve productivity.

Hoist procurement, in short, is a crucial decision for the project. Getting it wrong can be an expensive business, because deciding subsequently to replace a hoist or to add more hoists – or to create more holes in the building – will increase the project costs. When procuring the hoists, the logistician should also remember to attend to the ancillary items needed to make them operational – the scaffolding, the base to put the mast, the run-offs or ramps. Also, it should be borne in mind that they will be installed at a reasonably early stage, perhaps before everything needed is in place. As with other forms of plant, they won't work unless you ensure that you have enough power to operate them.

It must also be remembered that, as mechanical devices, hoists will need considerable maintenance. In the first instance, they need to be installed properly, with all parts tested and checked. It's important to buy hoists from a reputable hoist supplier, and to schedule regular maintenance. The fact that a hoist will need to be down for at least four hours' maintenance each month is another consideration that should be allowed for at the programming stage. Scheduling these checks is a standard part of logistics management, but it must also be understood that hoists can and do break down despite regular services. Ice and frost, for instance, can be a problem, because the metal components can fatigue in harsh weather. Therefore, the logistician should be aware of the terms of the Service Level Agreement with the hoist supplier. Of course, the higher the level of service provision procured, the more expensive it will be. It can be very tempting to hope for the best and settle for an agreement where there will be a response within 24 hours, but in practice this can result in the loss of more than a day's work. The exact level of service procured should, therefore, be carefully considered and debated with the team.

It makes sense for the hoist operators to form part of the logistics team. Hoist operating is not, perhaps, the most exciting job in the world. The conditions are noisy and can also be chilly, and there will be people constantly after your attention. This means that it is unlikely to appeal to many workers in the longer term, and is therefore an ideal job to be rotated around a larger logistics team. Hoist operators should also be trained. This is currently best practice rather than a legal requirement, although certificates will now be expected on better sites, such as those under the auspices of members of the Major Contractors Group. The role is not especially complicated, but the training is extremely important from a health and safety point of view. When all goes well, operating a hoist is merely a matter of knowing how to press a button or move a handle. If the hoist develops a fault, however, the training comes into its own. Regular refreshers should be given to all operators so that they know what to do if the hoist breaks and how to get down safely if they are stranded. Hoists are vulnerable to wind, so the operators also need to know when to use their judgement and ask their supervisor to suspend the service.

It should be remembered that there is the option of driverless hoists, but such facilities can be open to abuse and vandalism. Where used, there is a tendency to have a driverless and a manned hoist side by side. In Japan, by contrast, there are pay-as-you-go driverless hoists. Each trade contractor has a code and is charged for their use of that hoist. In the UK, this seems unlikely to become the predominant modus operandi for the foreseeable future, but it is certainly another option to consider when planning requirements.

The scope and potential of common user plant

It would take far too long to discuss all the different types of plant and hoists available and, in any case, the plant and equipment available for construction projects is improving all the time. However, it is worth saying a little about getting the principles right. As with tower cranes and hoists, the trick is to procure the right equipment and ancillary items in the first place, and to ensure that the practicalities of deploying them have been properly thought through.

It is also suggested that, whilst some items are established by custom and necessity as common user plant, other types of plant could be usefully put within this scope.

In particular, the author believes that one useful change, which the industry has not, as yet, taken up in any significant numbers, would be to include mobile elevating work platforms (MEWPs) within the category of common user plant. MEWPs are becoming more and more popular, and come in ever-increasing varieties. They can be more lightweight and agile, or bigger and more robust, than ever before. This means that they can reach greater heights, and are especially popular with mechanical and electrical contractors, who spend much of their time fixing things to ceilings.

At the time of writing, there will, on any reasonably sizable construction site, be a plethora of MEWPs, generally standing idle and parked in someone's way. For the individual contractors, it's often easier to hire them for the whole fixed period of a contract, since they are all responsible for the procurement of their own MEWPs so that they are on hand whenever required.

This means that the modern construction site can take on a striking resemblance to an MEWP car park, especially if that site happens to be building a shopping centre, the kind of project where they will be used extensively.

MEWPs standing idle are, quite clearly, wasting money as well as being an extra logistical inconvenience. If they were added to the remit of common user plant, assessing the number of MEWPs required for the entire project would not be difficult. The logistician would simply need to ask all the trades to give an indication of how many they would need. If MEWPs were therefore under the control of the principal contractor, there would be fewer on site, and they could be hired to the contractors on an as-used basis. No money would be needed for the transaction, as sales could be recorded electronically.

The logistics team could take care of the delivery and removal of all MEWPs, rather than relying on the trades to bring their own. An added advantage would be that it would eliminate many unexpected deliveries, since when trade contractors have plant delivered they tend not to announce it since it's not part of the materials handling function. In an ideal world, MEWPs could be stored at CCs.

The same principle applies to zip-up towers on site, which are also not used efficiently on many sites, and are therefore sitting around for ages. Sometimes they are dismantled so that they take up less room, but this means that parts can be lost or misappropriated.

In a worst-case scenario, a site can have 25 mobile towers with bits missing, some or which will have been hired. This problem means that the vast majority get skipped at the end of each job, and so losing them is often built into the overall cost.

The reasons for adding such items to the list of common user plant, therefore, would seem compelling enough to prompt action. In practice, however, there is a concern that this will increase the risks to the principal contractor. If someone requests plant at a time when it's already in use, that could create grounds for a claim. Nevertheless, considering that current utilisation of such plant probably runs at less than 30%, it would be worth trying to solve these issues. If nothing else, it would save the problems that often come at the very end of a project, when the logistics manager is trying to locate the owners or leasers of a dozen different items, in the hope of getting them off-site in time for the final handover.

Couriers

Couriers can cause complications. The eponymous 'white van man' and the motorcycle courier can be regular visitors to site. They will be bringing small consumable items, such as screws, mastics, drills or other handheld equipment. The courier will not have a delivery slot in the loading bay, and so may park inappropriately – either on a pavement or on double-yellow lines – whilst they wander around the site for 20 minutes in search of the person who ordered the screws. That person may be several tiers down in the supply chain. The courier will be asking for John Smith. No one will have heard of John Smith, who has gone for lunch anyway. This process can take some time. In the meantime, the van is in the way.

This will have been initiated by the subcontractor concerned, but the perception is that the traffic marshals within the logistics team are responsible for managing the situation. This can create problems, so the shrewd logistician will want to find a method for dealing with courier deliveries. You can't stop couriers coming to site, since requests not to use them are likely to be ignored, and even the most senior logistician has no control over the Road Traffic Act, so you want to deal with them as quickly and efficiently as possible.

There are several potential solutions. If the site is using a CC, then clearly it makes sense to have couriers deliver there. On an ordinary site, another alternative would be to make receiving such deliveries an extra part of the security officer's remit. In this case, they will need to be given clear responsibility and facilities. If you have a parcels receiving area, you will need buy-in from the trade contractors, since usage of the facility will need to be at the contractors' own risk in case of lost or damaged items, so that the security officer can sign for packages.

The ideal situation would be to have a secure store, so that the security officer can sign for items, log them and store them in an orderly manner, logging notes if a package has been damaged. This might also be countersigned by the courier. The security officer can then use a mobile phone to text the trade contractor to alert them of the parcel's arrival. To ensure that this facility doesn't get used as a storeroom by the trades, it is wise to give each delivery a definite shelf life. The contractors need to understand that they have 48 hours before their parcel will be disposed of; otherwise, the facility will end up swamped by thousands of parcels.

CCTV cameras in reception can provide an extra record of each delivery and its condition. The store will need to be kept locked and robustly secure, since some items will be expensive. This, however, is reasonably simple to ensure, providing a straightforward solution to what can otherwise be a real problem for everybody involved.

Alternative forms of transport

Much is spoken about the benefits and importance of finding alternative forms of transport to site, and in recent years there have been a number of suggestions in press articles about how using water and rail would be more sustainable approaches. Whilst this is a laudable ambition, the realities of life dictate that for the vast majority of projects it is hard to see how this ambition could be translated into practice.

Often, the adoption of alternatives is only partially achievable even on the biggest projects, and at the time of writing the only project where such practices are really significant is the Olympics of 2012. Valuable lessons will undoubtedly be learnt from the Olympic experience, but it is important that the achievements of this particular project be placed in context.

The special nature of the site plays a great part in the Olympic achievements. Not only is it a mega-project in terms of scale, superseding even Heathrow's fifth terminal in terms of its size, it is faced with a situation where the barriers to good logistics and efficient construction are the same as the solutions. In particular, a number of rail lines from Stratford will be restored, creating freight sidings by the sites.

The Olympic project would be a mega-project in its own right, but it is also inextricably linked to the Stratford city project, which is creating a shopping centre and office buildings. The third piece of the jigsaw is the athletes' village. These three projects are essentially one logistical challenge. As well as being sited in an area with scope for the resurrection of its existing infrastructure (which, as we shall see, applies to both rail and waterways), there has been the added driver that so much construction activity would otherwise have blocked the A12 and surrounding road network.

Therefore, this project had both the opportunity and the incentive. It would be difficult for a normal construction site, or even most super construction sites, to use rail and water transport unless this infrastructure already existed. Although the Olympic Development Authority has to be admired for its rail and water ambitions (50% of materials to be moved by these means) and for its success, it must be remembered that it has had huge advantages in terms of the existing infrastructure.

There was a rail head there already, and suppliers of aggregates are used to transporting their goods by rail. The difference on this project is that the rail head at Bow East allows them to get their goods much closer to the point of use. It could be argued that when rail is used on a project without such geographical benefits there is still a significant amount of lorry miles when the materials are unloaded. With the Olympics, in contrast, it's impossible to ignore the efficiencies offered by the rail network. One train can transport the equivalent of perhaps 75 lorry loads. Where there is an existing rail line, this is very viable, but in the wider rail network most freight moves out of hours and passenger trains are the priority. This means that if there is a hold-up for track repairs freight often drops off the list and can be sat in sidings waiting for a new slot for transportation. In short, rail does have great advantages, and can create significant carbon reductions in comparison to road transport, but it may be both impractical and expensive unless you're working on a project of Olympic scale.

A waterway network was also purchased as part of the land acquisition of the Olympic park, although it is not in the park itself. Still, the waterway needed a huge spend to make it navigable, and there is a five-metre fluctuation in depth, depending on the tides. The spend of perhaps £25 million that was needed to make the waterways viable was largely driven by the need to create an aesthetically pleasing area as part of the Olympic legacy. No one wants to see shopping trolleys floating in canals when the world comes to London. Even so, this spend would be far beyond almost any other project, even if there was a waterway available to be reclaimed. The fact that barges can now play a part in delivering aggregates means that one load can replace 8.5 lorries and can create substantial reductions in CO_2. Water transportation is even greener than rail.

However, water is slow, and this particular project benefits from a big aggregates depot on the Thames river at Tilbury dock, which is accessible for barges. Water is not suitable for moving smaller items, and is only successful over short distances. At the time of writing, not much material has come into the project by barge, and the forecast is that not much will do so over the life of the project since the aggregates are coming in efficiently by train, which is more effective.

Sustainability concerns and the need to use this waterway makes it likely that barges will be used for the removal of waste from the Olympic site. Landfill sites are situated on the Thames, so that will be an efficient strategy. Some recycling plants and waste-to-energy plants are also sited on the Thames.

Therefore, the Olympic model works after a fashion, but, whilst there has been much rhetoric about the desirability of its approaches, it's important to understand that not every future construction project would wish, or be able, to replicate it. A good logistics manager has to consider all options, and should a project be situated by a river or railway, these options should certainly be considered. If there is no existing infrastructure, however, it will probably be practically impossible to implement such as strategy. The Olympics have demonstrated that alternative forms of transport are good for CO_2 emissions, and they are utilising logistics to regulate vehicle flow into site. The author certainly wouldn't want to discourage an innovative logistics manager from finding alternative transport solutions, but feels that a note of realism about what is achievable in the life of a more typical project is an important contribution to the sustainability debate.

Distribution

Whilst it is important to have a delivery management system to provide a control on deliveries, whether the deliveries are dropped off directly at the site or via a CC, a lot of the snarl-up starts to occur once the materials are at the loading bay.

Depending on the nature of the site, it may be possible to allocate certain areas for temporary storage. The author believes that this should be discouraged wherever possible, since it leaves the materials vulnerable to damage, but on some constrained sites it may be the only solution. Such areas should be clearly marked out and have signs up to say which space is for which contractor, but, with the best will in the world, all too often people will still be trying to make their space bigger. It's not entirely acceptable as a solution, and just-in-time deliveries are to be preferred.

Nevertheless, whatever the system in place, somebody has to unload materials, take them to site and either place them in the lay-down/storage area or distribute them direct to the workface for the trades to use. As we mentioned earlier, using the tradespeople to do this task is counterproductive, since it is not a good use of their time and can have a negative effect on morale. Employing casual labour can also be unwise. Still, these are the two most commonly used options. The distribution of materials around site is a vital function, but it is quite commonly overlooked by the principal contractor. If it is not overlooked, responsibility tends to be passed to each package contractor.

When distribution is efficient, deliveries are unloaded in a timely manner and go straight to the right place, as demonstrated in Figure 5.4. Worst practice, which is all too prevalent in the industry, is an entirely different operation. Materials are unloaded and then left either on the pavement or somewhere else utterly unsuitable, to be taken in piecemeal as required. This is bad for the industry's image, causes congestion in the unloading area and vastly increases the chance of materials being damaged, stolen or going to the wrong place. The problem of materials going to the wrong place is far from the trivial matter that it may appear. In a 10-storey building – which would be far from exceptional as a construction project – if materials go to the wrong floor, they are, to all intents and purposes, irredeemably lost. The workers who needed the delivery may spend some time searching for them, but they are unlikely to comb the whole building. If the load does not turn up on an adjacent floor, they will probably simply reorder what they need. This makes us appreciate how easy it is to accrue a huge amount of wastage on a very average construction project. It has been reported that on projects with no specialist logistics involvement 'an average of 10.1% of the working day, or 49 minutes per person per day, was spent collecting, or waiting for, materials' (Hawkins 2008).

Besides this, of course, when materials are delivered to the wrong place they are likely to be in someone else's way, which means they may

Figure 5.4 Efficient unload of materials from a consolidation centre delivery vehicle at the Skanska Barts project

get damaged, especially if the workers impeded by the load attempt to move it without the right equipment. It's quite likely to go straight to landfill.

In manufacturing, components will usually reach the production line with such efficiency that it's literally impossible for the operatives to fit the wrong part: red cars go through the line on a Tuesday; blue cars on a Thursday. Getting the wrong colour fitted is impossible, because the supplies are engineered to reflect the production schedule. This is impossible in a construction environment. The project is delivering something far more complex, and it cannot be built on a production line. This makes a professional approach to deliveries even more important. For instance, duct work can all look the same but trying to fit a part built for slightly different size spaces will eat away time before the mistake is discovered.

This is just one of the reasons why it makes sense to have a dedicated materials handling team to manage the unloading and distribution of materials. When one group take responsibility for getting the right materials to the right workface, many of the irritations and inefficiencies that are now commonplace start to melt away. One of the most important benefits is that the trades will start to enjoy coming to work, since the right materials are there for them to use, and this enhances quality measurably.

When a delivery throws the schedule by arriving late, you can either take a coordinated approach to fixing the problem or rely on survival of the fittest. Despite the potential efficiencies of the former course of action, construction tends to opt for the more Darwinian solution, because it devolves risk onto the trade contractors. If the trades are responsible for their own equipment, it's their fault if things get broken in transit. Instead of managing the risk by using a professional team, the risk gets passed around, so materials are ultimately handled by people who are either too skilled (trades) or not skilled enough (casual labour) to produce the best performance.

Furthermore, since this problem is devolved, there is usually no concerted attempt to programme works in a way that would facilitate the smooth movement of materials. Even the most casual observer might deduce that it would make sense to put any roads in first. Building any loading bay that will comprise part of the finished product at an early stage would also seem to make sense. Often, however, it doesn't happen. On a site, internal access ways into huge buildings around the site are generally not finished until the end of a job.

This situation also seems to have an impact on the quality of materials handling equipment. The uneven surfaces of a site do not easily lend themselves to the utilisation of wheeled equipment. Any lifting equipment that is available thus becomes more difficult to use. Logistics in the manufacturing and retail sectors can draw on state-of-the-art mechanical lifting devices, whilst construction still relies on pulley systems that are not so different from the lift devices of the ancient world. The pyramid builders would recognise many of our current technologies.

In construction, there are many inhibitors to investment in decent equipment. Not only is there the contractual devolvement and the uneven terrain, there is also the problem that a project-based industry is prone to ending up with expensive equipment lying dormant at the end of a job. Continuity in construction is uneven at the best of times. In a recession, the arguments against investment are even stronger.

This leads to a make-do-and-mend approach. When materials are not moved using brute force, they are moved using whatever's available. Wheelie bins intended for waste removal are pressed into service as makeshift trolleys. They are clearly not designed for the job, and this is a potentially serious health and safety issue. Even where there is no actual incident, the ergonomic cost to workers is a harsh one. It has been found that, whilst older construction workers often take pride in their resilience and under-report health problems, many are struggling with chronic problems such as backache (Collier 2007). Over decades of strain and manual lifting, workers adapt to a life spent, quite literally, in pain. Not only is this a tragic human outcome, where thousands live with a poor quality of life due to current working practices, it is also a serious impediment to industry attempts at diversity. The pool of workers construction draws on is divided into two groups. There are those who may get permanently damaged by the loads they lift, and there are those without the strength to lift them in the first place. No worker benefits from this situation.

There are also dangers involved in actually unloading deliveries – another argument in favour of using a properly trained team with the right equipment and handling techniques. Unloading by crane, for instance, requires the knowledge to sling a load correctly using straps, strops and chains. Often, someone needs to climb onto the back of a flatbed lorry, which will be five feet off the ground and devoid of fall protection, to secure the load. As the load leaves the ground, depending on a number of variables, such as wind and the camber of the road, it may start to swing. The slinger has, quite literally, nowhere to go. They may be reduced to trying to shoulder-charge the load. For this reason, people fall off trucks far more than might be expected.

If the issue is understood, however, there are many ways of unloading more safely. Crash mats round a truck are a sensible precaution, as are fall-arrest systems, perhaps with a gantry above the loading bay. Many interventions can have a role to play, but many sites have no systems in place, because of the ad hoc nature of their approach to materials handling. If you have a trained and dedicated team and a structured approach, on the other hand, you reduce the chance of injuries occurring.

It can be hard to write method statements for unloading operations, simply because you do not always know how a lorry has been loaded. Some loads are usually only accessible from the back; some are only accessible from the side. It should also be remembered that a load might have moved in transit and become unstable. There are a number of factors to take into account, which doesn't allow you to plan precisely. You may be expecting a delivery on pallets, only to find it's actually loose-loaded. More and more sites are actually banning people from climbing onto trucks. This means that manufacturers and logisticians will need to find new ways of working.

There are a variety of things to consider, which the author believes constitute another argument in favour of the CC. The right equipment can be used to move the right materials. Also, the actual deliveries to site can be packed to make them easy to unload. When goods are travelling overland across a continent, the load will of course be packed to get maximum value from the truck. That's an argument in favour of specialist unloading in a warehouse environment, so that loads can be inspected and rationalised before they are actually sent to site.

After all, since materials are the lifeblood of any site, and must be dealt with on a daily basis, why not do so in the most efficient way possible?

Batching plants or concrete deliveries?

Construction uses an inordinate amount of ready-mixed concrete, a material which comes in many varieties and has myriad uses. The technical intricacies of concrete are beyond the remit of this book, but the logistical options for obtaining it must be mentioned.

There are two ways of getting concrete to a project. Either it can be delivered ready-mixed in trucks or it can be made on site using a batching plant. The constituent parts of concrete are cement dust, sand or aggregates and water. Setting up a batching plant is fairly expensive. On top of this, such a facility requires a certain amount of land for plant and for the storage of aggregates.

Given the cost, when might such a facility be considered? Essentially, there are two possible reasons. One is the economies of scale that can be derived if there is to be a very high use of concrete on a very large project. If the construction calls for more than 100,000 cubic metres, a batching plant may be the cheapest option. Setting up a plant is unlikely to be economically viable for a project using less that 50,000 cubic metres. If the project will involve pouring 50,000–100,000 cubic metres, there may be marginal cost savings, but the decision to build one will be largely affected by the other major driver for on-site production: certainty and security of supply during a major pour.

Incidents such as road traffic accidents, and any other circumstances that could delay the arrival of a ready-mix truck, can be fatal to concrete pouring. When one remembers that a large pour might need the equivalent of 30 or 40 lorry loads, and considers the likelihood of at least some of these vehicles being delayed, the attractions of an on-site batching plant become clear. Deliberately engineered delays, such as the possibility of protesters lying down in front of trucks on a controversial project, might also affect the decision.

These are the two major drivers, although the situation of the project might also be a factor. If there is no local source of ready-mix, a batching plant might become attractive on a project that would not otherwise use one.

There are two possible routes to setting up a batching plant. In theory, the logistician could go out and procure all the necessary kit and equipment to set it up. In practice, however, this is unlikely. The second option, which would be to ask a national concrete supplier to set up a dedicated plant, is usually the preferred route. The site provides the land; the specialist provides the kit, equipment and expertise. This is usually more efficient, because the specialist will have, or know where to buy, the right equipment.

Typically, the cost of this type of agreement will be largely dictated by how much concrete you intended to use. There will probably be a contractual minimum and maximum spend specified. If these quantities are large enough, a generous discount on set-up costs may be given in return for a guarantee that the specialist can provide all the concrete.

Another ancillary driver for opting for a batching part, particularly from the logistician's perspective, is that it will significantly reduce the number of concrete-related deliveries to site, perhaps by as much as 40%. This capacity can be either assigned to other loads or used as a measure of environmental performance. Reducing the number of lorries arriving on site will of course reduce the number of vehicles leaving site requiring wheel-washing prior to them joining public roads.

Still, aggregate stockpiles will still have to be delivered, and perhaps 5000 square metres of land will be needed to store them. The plants are quite compact, but the equipment will still need to go somewhere. Batching plants also need a ready supply of water to create the mix, and somewhere to dispose of the grey water.

When siting the plant, consider opportunities to pump the concrete through a wet hopper to the point of use, rather than having it run 500 metres in a truck. Without a hopper, trucks will still be going backwards and forwards, albeit over much shorter distances.

To set up a batching plant, it may be necessary to get planning consent. At the very least, there will need to be a conversation with the relevant Local Authority. Batching plants are not particularly noisy, but the loading of aggregates will create some noise. Neighbours who are sited downwind of the plant may also experience nuisance from the cement dust.

Setting up a batching plant from scratch requires a lead time of 8–10 weeks. The exact time needed will be dictated by the type of concrete being produced. The feasibility of setting up a batching plant should be considered during data capture and analysis. When undertaking demand planning for vehicle movements on site or through the local network, ready-mix deliveries, if used, need to be taken into account.

The costs dictate that there will not be a batching plant unless there's a compelling reason for having one as part of the overarching strategy. Drivers such as programme certainty, security of delivery and the environmental benefits will have to be weighed up against the cost in order to ascertain the business case. Either way, it forms part of the overall strategy.

A lot of ready-mix trucks are operated by owner/drivers, as self-employed workers who are accountable to nothing except the demands of the next concrete pour. When multiple ready-mix trucks are queued up, as they can be to guarantee a constant supply on a large pour, this can create the most demanding situations faced by traffic marshals. The queue can cause congestion on the local road network. It's practically impossible to move this crucial supply line, even if the drivers were minded to move, because of the demands of the project, so the logistician must be aware that the situation may incur the wrath of the police. The traffic marshals also need to be aware that a hard-pressed owner/driver may not be entirely amenable to requests to remove part-loads of concrete that have proved to be surplus to requirements, since their services will often be urgently required on their next assignment. This can lead to surplus product simply being dumped *in situ*. When removing it, it's important to remember that this is a material with inherent dangers, because of the chemicals involved.

Barcodes and radio-frequency identification

Barcodes and radio-frequency identification (RFID) are the two technologies generally used for tracking kit and materials.

Barcoding is commonly used on retail products. The construction industry has tried to embrace it, but it has generally been unsuccessful. On the face of it, this is somewhat surprising. After all, the big DIY retailers do it on many of the same products. Nevertheless, it's very difficult to achieve in construction. All the issues, discussed elsewhere in this book, that make it difficult to create a coherent construction logistics function – the usual forms of contract, the fragmented supply chain and the reluctance on the part of architects to specify products – also make it almost impossible to barcode items at source.

In theory, it is possible to barcode materials at a construction CC, but by the time the items reach that stage there is very little point in doing it. Materials will already be tracked by a warehouse management system; so, by the time you've created barcodes, you might as well just make an easily comprehended label. Barcoding wouldn't add huge efficiency at that point.

Another issue is that barcode systems will mark larger items individually, and mark quantities in bulk. Therefore, a system in construction would need separate barcodes for bulk deliveries, cartons and individual products. In short, such systems fall into the 'too difficult' box, as they would need to be implemented on a one-off basis every time. In the volume-housing sector, which does commonly use standard products, it might be feasible. In commercial and industrial construction, however, it is difficult to see how it could add value.

Many of the same issues apply for RFID tagging, since this is essentially a more sophisticated barcode. RFID identifies products or pieces of kit using technology that can be tracked by satellite and read from a distance by handheld scanners. The advantage of RFID is that more information can be embedded than would be possible with an ordinary barcode. RFID can record when an item was fitted. It can track the dates when the item should be maintained, and when it should be replaced. Clearly, this is of potential benefit to construction. Still, however, challenges to getting the right information on the right bit of kit are created by the contractual and cultural inhibitors previously mentioned in relation to barcodes.

RFID is also hugely expensive to set up. That's okay if you're a major retailer, perhaps selling 15,000 units of the same product, but it's rather different when you're trying to tag bespoke construction materials for a one-off project.

For these reasons, materials tracking remains an ambition rather than a reality at the time of writing. That situation seems likely to endure unless there is a radical change in construction's core processes. In contrast, tracking plant is a rather easier conundrum to solve. The technology used for satellite navigation and GPS can be easily adapted to fit construction plant. Construction utilises many different specialist – and expensive – items of plant, from MEWPs to earth-movers. In such cases, the technology is generally fitted for security purposes rather than as a form of logistics.

Consolidation centres

CCs are an important emerging approach to construction logistics, and the rationale and methodology for operating them is covered in detail in Chapter 11. Nevertheless, it would be impossible to write about the logistics of construction materials handling here without giving them some mention. Figure 5.5 shows a CC delivery vehicle being loaded with the exact quantity of materials required by the project it is supporting.

The core purpose of a CC is to create a buffer to mitigate some of the problems discussed above, and to bring efficiency to supply chain logistics. Since there are inefficiencies both in getting materials to a site and in moving them around the site, having a CC as a buffer between the two functions can be enormously helpful. A useful analogy is to visualise trying to pour water into a bottle. It you fill it straight from the tap, water usually splashes everywhere, however careful you are to try to control the rate of supply. If you use a funnel, on the other

Figure 5.5 A consolidation centre delivery vehicle being loaded with the exact quantity of materials required by the project it is supporting

hand, the task becomes infinitely easier. The funnel regulates and guides the flow, and thus prevents damage, wastage and inefficiency. This is exactly what a CC can do for the supply of materials to a construction project.

It has been demonstrated that construction CCs can achieve substantial savings and efficiencies for a project. (See, for instance, Hawkins 2008.) Therefore, a reader might reasonably ask, why isn't everyone using them?

The answer to that seems to be cultural, and reflects both contractual complexities and ingrained approaches to the practice of construction. The norms are difficult to challenge – in the words of Mossman (2008): 'culture eats change for lunch'.

To begin with, there may be a perception that a CC is in some way planning for failure. It is built around a system of double-handling. That won't seem attractive to anyone in denial of the fact that poly-handling is an industry norm. Adopting this method involves the, sometimes uncomfortable, recognition of suboptimal elements within the status quo. The approach should be seen as a rational solution to the inherent complexities of the construction process, and not as an admission of failure. There are compelling practical and financial arguments in favour of consolidation.

Cultural inhibitors, however, can be both deep-rooted and subtle. The reality of life on site encompasses an abiding cultural norm where people are expected to overcome obstacles at the behest of the person shouting loudest. 'Just do it' is a common enough refrain, but is not a command that takes any account of the inherent problems involved in the detail of various aspects of delivery. The result, too often, is that no one actually sits down and works out that reaching a target might actually be physically impossible within the given practical constraints. This can mean that suppliers promise what they can't deliver, and everyone becomes complicit in the resultant failure. Often, every stop has to then be pulled out during the last days of the project, when the spend will ratchet up to compensate for delays in the programme. When the project is – magically, miraculously – finally delivered on time via a desperate and inefficient last-minute effort, it does indeed represent an impressive achievement. Perhaps only a week before, it had looked impossible. Now, there is a building and a handover.

This has long been the way that construction creates its heroes, a phenomenon also reported by Mossman (2008). Within the industry, managers who pull a project back from the brink are lionised. Managers who bring in projects that never got anywhere near to the brink, in the main, are not. After all, consistency and efficiency are hardly the most exciting virtues. Their peers are not impressed by the timely delivery of a job where 'nothing went wrong'. The budgetary implications of a project that nearly went very wrong indeed tend, within this industry's folklore, to be glossed over somewhat. After all, rescuing impossible situations is the stuff of romance, and romance never counts the cost of heroic achievements.

The ingrained tendency of the construction industry to fire-fight using an amalgam of machismo and magical thinking (if I want it to happen enough, it will happen, especially if I shout loudly, runs the thought) is the antithesis of the orderly calm and predictability offered by the CC. It may be that, without a fundamental change in values on the ground, this solution will never be embraced by the industry at large. However, if the industry could learn to plan for what is likely to happen, rather than what should happen on paper or in the realm of wishful thinking, there would be a whole raft of benefits and values to industry performance. At a CC, the logistician will ask the trade contractors on Friday what materials they require on Monday and Tuesday. Their requirements are met exactly within 15 minutes of the agreed delivery time. Everything is sorted into work packs. Thus, the process of getting materials to site is smooth. The centre builds in space and capacity – the things that are at such a premium on the average site. Along with increased efficiency comes a better financial and environmental performance. Savings commonly amount to 2.5% of the total project spend.

To operate well, the centre needs to work hand in glove with the materials delivery team, which is going to unload the materials efficiently and take them to the point of use. That way, you get the right materials in the right place at the right time. The importance of this cannot be over stated.

The philosophy behind CCs is, clearly, informed by the theory of lean construction. That said, the author believes it is possible to be too lean. Lean is a useful tool, but construction is still a very human process. Rescuing derailed projects at the eleventh hour has been an industry norm for too long, but that observation should in no way be taken as belittling the fragmentation and complexity successfully managed within the construction process. Even with the best-laid plans, things can and will go wrong. For that reason, the art of the CC is to provide the efficiencies of lean, with just enough extra capacity to act as a buffer to the unforeseen. It's a balancing act – enabling the smooth execution of Plan A whilst creating the capacity to implement Plan B in the event of a crisis – but at its best a CC can manage a very high percentage of crises out of the equation before they ever happen. This leaves the construction professionals free to do what they do best – create state-of-the-art buildings that provide us with both the artistic landmarks of our skyline and the creature comforts of modern life. One might imagine that every CC in the country would be filled to capacity, with the industry crying out for more. At the time of writing, this is not the case. Whether this is really due to an abiding addiction to the adrenalin rush of crisis management or to another reason for why the industry is so reluctant to save so much time and money is a matter we could all be usefully debating.

References

Collier C (2007) Safety through the ages? *Construction Information Quarterly* **9**(3): 142.

Construction Products Association (2009) Introduction to the Construction Products Association, http://www.constructionproducts.org.uk/, [accessed 29th August 2009].

Hawkins G (2008) *A Study of Materials Logistics on Three Office Developments in the City of London. Final report 19817/3*. BSRIA, Bracknell.

Holt A (2001) *Principles of Construction Safety*. Blackwell, Oxford.

Mossman A (2008) 'More than materials: Managing what's needed to create value in construction'. Paper for the Second European Conference on Construction Logistics, ECCL, Dortmund, May 2008.

Transport for London (2007) *London Construction Consolidation Centre Interim Report, May 2007*, http://www.tfl.gov.uk/microsites/freight/documents/publications/LCCC-interim-report-may-07.pdf, [accessed 12th January 2010].

Chapter 6
Transport and Communications

Traffic management is one of the lynchpins of any logistics plan. A number of elements must be carefully considered. The movements of vehicles, pedestrians and plant must all be planned, because if any one element fails to work it can temporarily derail the project. Arguably, traffic management can have the greatest overall impact on the success of a project.

Clearly, effective traffic management, as with most areas of management, will be largely down to good communication. These are allied areas – indeed, when we remember that roads, as much as telephones, are part of our communications network, we start to realise how closely transport and communications are related. Where transportation moves people, plant and materials, communications move data and ideas. Both need to work in an effective and timely manner for the project to run smoothly.

Introduction to traffic management

As with so many aspects of logistics management in construction, there is no one-size-fits-all solution to traffic management, but the key to creating a successful plan is knowing what you will need to manage.

Systematic data capture needs to start well before the project commences. The logistics manager study several interrelated factors, particularly relating to materials. Types and dimensions of material must be considered in conjunction with data on their likely volumes and flow.

The logistician must ensure that the right capacity will be achieved in terms of hoists and other lifting equipment, since efficient materials handling cannot be considered independently from efficient traffic management. Then, the logistician can start to create detailed plans about how deliveries will be processed. Indeed, there is a sense that the pre-construction phase is more intense than the actual construction for the logistician in charge of transport. So long as sensible plans have been created, which take into account the real constraints of the site, the construction phase can be largely a routine management exercise. If, by contrast, the correct plans are not in place, or they will not work in practice, the construction phase can be very challenging indeed!

Plans need to become progressively more detailed as the construction phase gets closer. Initially, there will merely be indicative figures around the tonnage which will probably have to be handled in any given month. As the start of the project draws near, fine tuning and day-to-day planning of both materials requirements and traffic movements become a necessity.

To achieve them, the logistician needs a detailed and clearly delineated construction programme, which he or she will have studied in depth. All the phases of the programme, from demolition to final fix, need to be thought through. Of course, the nature of the project, its size and any special requirements will all need to be taken into account. This is a fairly detailed exercise in scientific planning, but logistics is also, in a sense, an art. Only experience really helps the logistics manager to anticipate all the particular requirements which will be created by the different facets of each, unique, project.

Some issues, nevertheless, must be accounted for on most, if not all, projects, and the following sections identify some of the main points to consider.

Managing construction traffic on the highways

First, it is important to understand the routes into and out of the site. If the site under consideration is at the centre of a city, the routes in from the motorway network need to be mapped out. If it's a greenfield site, on the other hand, issues like ensuring that the route avoids passing village schools become more important.

In either instance, the goal is to ensure that the planning of vehicle movements is done in a way which minimises the impact of the extra traffic on the local community.

Having researched all the likely constraints and collected data on both the volume and the degree of difficulty of deliveries (which will depend largely on the size of vehicles and their likely turning circles), the logistics manager needs to start to physically draw out routes to site, starting from the major trunk roads. The plan will be informed by knowledge of where the bulk of the materials will come from. For UK projects, for instance, a significant quantity of materials will generally arrive from the Midlands and the north-west. There is also likely to be material arriving from other countries, which means, in the UK at least, that the logistics manager will need to concentrate particularly on working out efficient routes from the major ports.

Another issue that must be borne in mind is that some oversized loads may require special measures. These measures could include police escorts, and/or the scheduling of movement during less busy periods, especially avoiding the rush hour. Another increasingly prevalent factor is the creation of exclusion zones in larger cities. These zones exclude HGVs, unless they are compliant with emissions standards, and the wrong vehicle entering the area will attract a significant fine.

Likewise, it is likely that the Local Authority or police will have preferred routes by which the site should be accessed, in order to avoid travel near schools, along popular roads or under or over bridges with limited height or weight-bearing capacity.

The logistics manager will need to think about all the various external activities which might affect transportation to the site. Since these are, by definition, out of the logistician's control, there is only a limited extent to which they can be managed. It is important to liaise with the relevant authorities to determine when activities such as roadworks might have an impact, but it must also be recognised that it will be impossible to guarantee timely information on the activities of, for instance, utilities companies. Contingency plans should therefore be developed to mitigate the impact of restrictions to site access that these activities may entail.

When thinking about the impact of any disruption to the highways network, particular attention must be given to the most important group of road users of all: the emergency services. Their ability to access both the site and adjacent buildings should not be compromised at any time during the projects. Planning around their potential needs is especially important, and challenging, when working on space-constrained urban projects. However, it must always be uppermost in the logistics manager's mind that their access needs could be, quite literally, a matter of life and death, and therefore prioritised accordingly.

Managing access points

Getting large vehicles onto site safely and efficiently can be a particular problem. Lane closures, the temporary creation of one-way streets or the reversal of traffic flow may be necessary in some instances, which again will entail early liaison with the relevant authorities. The crucial point is to understand the turning circle of vehicles. The space required for this will largely dictate the set-up of access points within the site perimeter.

The logistics manager must also establish how many trucks will be arriving within the ambient (i.e. typical) traffic flow, and at what frequency. It's crucial to establish how long each vehicle will take to unload. This is one of the most difficult issues to plan, which is why construction logistics experts have become such strong advocates of construction consolidation centres, which are discussed in more detail in Chapter 11.

The time it takes to unload a vehicle will, of course, be determined to a great extent by how it was loaded prior to transit. It's important to liaise with suppliers on this, because simple details such as ensuring that a pallet is loaded the right way round can make a huge difference to unloading times. Unloading may require forklift access. It will certainly require robust risk management, and if people are to climb onto the back of a lorry, safety equipment should be in place.

Access to the correct unloading equipment is extremely important. It might be possible to unload an articulated lorry full of plasterboard in half an hour, but limitations in terms of hoists or carriages might mean it takes two hours to distribute the material around the site. Distribution will commonly constrain the speed of unload. Recognising and planning around these constraints is one of most important parts of the job, which is why it is vital for the logistics manager to understand the requirements of the design and the materials that will have to be delivered to create it.

The smooth running of materials delivery must always be a priority, because any failure in supply can have a domino effect: if one trade is not given the materials needed to work, everything else can back up as well. Even short delays can have a substantial impact: if delivery is taken after the relevant workers have left the site for the day, for instance, many other tasks may not be able to proceed.

Traffic management and materials delivery on larger sites will involve haul roads – temporary constructions which facilitate the movement of materials around the site (Figure 6.1). These have to be substantial in terms of their load bearing capacity, but are commonly simple construction of compacted stone, although on some larger sites the finished roads, at least to base coat finish, will be put in early. On larger sites, the construction and phasing of haul roads has to fit around activity, and may well need to be moved, perhaps more than once, during the life of the project. Some, but not all, haul roads will become permanent roads on the finished site.

Whilst deliveries will form a significant part of traffic to the site, the issue of how workers get there must also be considered. The actual mechanics of their ingress must be planned in detail. Bus stops may need to be moved, in which case early liaison with the Local Authority and the bus company is a prerequisite, Remembering that on larger projects upwards of 500 workers may need to arrive each morning, and that there may be only a short window of time for them all to clock in (see Chapter 8), the impact of this on both footpaths and the public transport network must be assessed.

Of course, the wise logistics manager will have dealt with issues such as unloading permits at a much earlier stage, because requests for permits

Figure 6.1 The traffic management plan created for the 1.25m square metre Stratford City site

will have to have been placed in good time to allow for processing. Likewise, many other forms of liaison, such as the moving or creation of bus stops, will need to have been in hand long before activity starts on site.

Car parking

On remote sites, the workers may need to be able to arrive by car, in which case the issue of parking needs will have to be considered. In an ideal world, the author believes, the parking of private vehicles on site wouldn't be permitted at any time. Aside from creating additional security risks, the priority for parking spaces can become contentious because it is perceived to be reflecting a hierarchy, space will be limited and extensive site parking, even if possible, will not be the most environmentally friendly way of getting workers to site. In the real world, it may be a practical necessity, but it will create issues that the logistician will need to manage.

Who is to receive priority parking will be one of the first issues to arise. Unless the project has a large site, space will be at a premium. It's best practice to have a system where authorisation for parking on site has to be signed by the trade contractors' site managers and passed up the line for approval. Everyone requesting a space will need to complete a form requiring their registration number and other vehicle details, and will need to submit it with a copy of their driving licence. The trade contractors' managers then pick the applications that they feel should be given one of their allocated spaces, and the logistician or site manager will sign it off. Passes for parking should be made specific to the cars in question, so that they can't be passed around or hawked. Site parking can be a significant added security risk, and this process of applications and allocation goes some way to manage that. Therefore, there should be a visible car park pass for security officers to check.

It's important to consider alternatives. On bigger sites, it may be necessary to have an off-site car park with an accompanying park-and-ride service. Public transport can be a good alternative, but on a larger site part of being a good neighbour will involve having discussions with the relevant transport authorities or providers. Super sites, such as the Olympics, may even have a dedicated contact. On this scale, there will need to be walkability studies and a careful analysis of volume flow. If there is only one railway station or bus stop, for example, 700 extra people are clearly going to have a significant impact. You may need to arrange extra buses to get people to work.

Whilst discussing commuting, it's also worth mentioning the need to consider the impact of commuters headed elsewhere. If your project is beside a mainline railway station or busy Tube line, hundreds, if not thousands, of people will be pouring out of the station and into the path of the construction workers and materials. In these situations, extra planning will be needed to ensure that the public are kept out of harm's way.

Managing haul roads

Managing the haul roads as the construction project unfolds can require much finessing. Signage, as is discussed in more detail later in this chapter, needs to be used at all access points, and at any other area where speed limits, directions or health and safety advice is needed. As far as possible, this should mimic standard signage and the provisions of the Road Traffic Act. Not only does this create less confusion for drivers, it also means that road users tend to treat this familiar signage with more respect than might be given to ad hoc notices around the site. It is likely that road lighting will be required, and often provisions for this will have to be made in areas where no mains power is available. The logistician therefore needs to be familiar with temporary lighting sets, and with how they should be serviced and maintained. Plant that is moving around a large site may need a fuel point or refilling station, which can be used to fuel generators for lighting as well.

The logistician should also ensure that as vehicles manoeuvre around the site there is absolute segregation between them and any pedestrians.

Sometimes, it is safest and most efficient to bus workers around the site, but otherwise clearly marked walkways and crossing points will be needed for those on foot. Traffic marshals should ensure that all vehicles go to the appropriate areas and that they don't park in areas where they could cause an obstruction. Vehicles can't be safely delivered without banksmen, so traffic marshals need to be qualified. They particularly need to understand how to sign and cone a traffic plan.

Wheel washes

In addition to traffic marshals, road sweepers have a vital role to play in health and safety and good housekeeping by ensuring that debris is not allowed to collect and form a hazard to traffic.

Nevertheless, during the substructure and superstructure works, it is likely that the site will become extremely muddy, so vehicles will need to be cleaned before they go back onto the highway. Therefore, a semi-permanent wheel wash or jet wash may be needed. This will, of course, require a significant amount of water, and so the logistics manager will need to think about where that will go after it's been used. It may be necessary to capture the dirty water and have it taken to a recognised tip with the ability to deal with contaminated water waste.

When a project is at the stage of digging foundations or piling, construction can become particularly messy work, especially when there is significant rainfall. When vehicles are exiting the site, they can be caked with mud. Not only would allowing them to leave like this be bad for the project's image but also the Highways Agency would not permit it. Mud on roads quickly becomes a safety issue, causing skidding. Debris becoming lodged between the double wheels of a lorry would constitute a significant safety risk when it started to work itself free. Because of these issues, Local Authorities will not allow works to commence until provisions have been made to deal with this.

If the size of the site permits it, a formal wheel wash machine is desirable. The vehicles drive on and keep their wheels turning whilst jets of water and brushes clean them. In reality, however, although these machines are improving, they won't always clean away all the mud and debris. There needs to be a visual check as well to see what's left between wheels – missing significant debris would be highly dangerous. In addition to the wheel wash, therefore, it's good practice to have a jet wash, so that these can be used if necessary after the visual check to take off the last bits of mud before the vehicle is allowed to exit onto the public highway.

The biggest drawback with both wheel and jet washes is that they will use up lots of water, and disposing of this water will need to be carefully managed. A wheel wash contains its waste and separates liquids from solids, but the water must be considered contaminated and disposed of accordingly. You can't put it in the drainage system, since it will be contaminated by soil. It needs to be disposed of in accordance with the hazardous waste regulations, so specialist contractors will need to be used for this. This, of course, creates significant cost, which it would be sensible to allow for in the overall budget. As obvious as this might be, it is frequently overlooked.

The responsibility for the installation and management of a wheel wash usually rests with the groundwork or piling contractor, but the logistics manager should double-check that this has been allowed for, and that the contractor in question understands the full extent of their duties. Construction scoping documents tend to rely on words like 'suitable'. A bloke with a hose does not constitute a 'suitable' wheel washing arrangement. The logistician must be prepared to spell this out where necessary.

An overview of communications

Communication in construction is notoriously difficult to manage. This was highlighted in the seminal Tavistock Report (Higgins and Jessop 1965), which pointed out the difficulties engendered by complex – and changing – relationships within the industry. In the words of later commentators: 'We were left to wonder how anyone managed to build given the difficulties identified. But continue to build we did' (Emmitt and Gorse 2003). Despite the complex nature of site communications, buildings still go up. And whilst there are misunderstandings, communication still, of necessity, takes place on a regular basis.

Communications for successful construction logistics need to happen at several levels. Most importantly, they often need to happen quickly, to pass on up-to-date information in a rapidly changing environment (Figure 6.2).

Whilst the importance of planning is a recurring theme in this book, plans will inevitably change. At the very least, there will be regular modifications. Communicating changes at short notice, providing accurate and detailed information regarding such adjustments, is therefore vital.

One of the most challenging characteristics of any construction site is the number of people involved in the decision-making process, which can lead to some confusion. Construction managers, the project manager, directors, quantity surveyors and trade contractors will all be part of the process, all adapting to change as the project progresses. The logistics manager could have up to 80 people to deal with on a daily basis, even on a standard construction project, and so the ability to be reactive and to adapt quickly is vital. Coordination and communication flow is vital for many reasons, but there will be a logistics dimension to almost all changes, whether the need is to close off a stair core, adapt materials or

Figure 6.2 The installation of the Pier 4 runway bridge at Gatwick Airport over the course of one weekend required up to the second communication

people flow or to change the logistics plan to ensure site safety in this rapidly changing environment.

The site logistics meeting

Formal meetings are one strand of the communications flow that helps keep all the different stakeholders up to speed with the changes. Sometimes those meetings need to be short and sharp; sometimes a longer and more involved exchange, possibly using a whiteboard so that notes can be taken, will be required. Typically, logistics meetings take place at around 3 p.m. By that time, all the managers involved will have a reasonable picture of what will have been achieved that day, and will thus be in a position to make detailed arrangements for the next day.

If things are going to schedule, this meeting may just formally confirm existing plans for the next day, but there might be variations where milestones have not been achieved. Alternatively, there might sometimes be variations because targets have been surpassed. Many factors can change. Suppliers sometimes have difficulties with the production or delivery of materials. There could be changes to road structures or neighbours could be carrying out an unexpected operation, which affects access. The weather could have prevented cranes and hoists from operating as expected. Even if the weather is fine, plant could be out of action through breakdowns.

Since so many things can change, it's vital to get all the key players into the meeting room. The actual negotiations which take place in that room can be rather like a City trading floor. People are giving out information and responding, and essentially bidding and negotiating for what they need – commonly delivery slots for materials – in a rapidly changing environment.

Arguably, other than the contractual meetings that involve large sums of money, these are the most important meetings that take place on a project. Supervisors and site managers are exchanging real-time information – and these are the people with a vested interest in getting things done.

Learning from the site

During the course of the day, the attentive logistics manager will find huge opportunities to capture information and data. Materials handlers, marshals and security guards – the core members of the logistician's team – will all be receiving information about progress and deliveries and finding out what is and isn't happening on the ground. This offers the logistics manager an invaluable grass-roots view of whether things are going well. That grass-roots knowledge (often processed and conveyed with a characteristic cynicism) can be very useful information if it is viewed as intelligence information rather than as carping.

Information can come in from the delivery management system. Permits to work are another important data source. Still, there is perhaps no substitute for the myriad conversations that the logistics manager will have with a range of people over the course of the day. Despite the growing range of communications technologies, 'management by walking about' still, in the author's view, remains the best way to give and receive information, and to have it challenged or verified.

Of course, that's not to say that we should discount technology, because it has certainly brought significant advantages, not least in terms of capturing and storing information. The speed at which information can now be transmitted to many people at the same time is also invaluable, despite the information overload which can be created by indiscriminate copying of emails to ever larger distribution lists!

Communication tools

In 1965, Gordon Moore, a co-founder of Intel, noticed that computer hardware power was doubling every 20 months (Moore 1965). This observation, now known as 'Moore's Law', has held up ever since, and has been stretched from its original definition to become a way of talking about exponential improvements in information communication technology (ICT). The rate at which technology is changing today means that many thoughts on modern ICT could be out of date before this book goes to print. Nonetheless, there are some tools that have become a common feature of the site, and are likely to remain within the logistician's armoury for some time to come.

Two-way radios are a long-standing technology, but they are still useful in communications between, for instance, crane drivers and banksmen. Hand signals are important, but radios have a part to play in the banking of plant. They are also useful for traffic marshals, security officers and logistics supervisors.

Radios work at the push of a button, rather than dialling a number. To be used to full effect, operators need to understand proper radio procedure – the use of terms such as 'roger' and 'over and out' is part of good radio discipline. This enables clear, precise information to be conveyed. That's why the police use radios in the same way, with a disciplined protocol for communicating.

Another important point to bear in mind is that, unlike with phones, everyone hears everything when two-way radios are used. This can be a real advantage in an emergency, and day-to-day it can be another useful source of information to the logistics manager. However, there is certainly a downside to this – when mistakes happen, they can be announced to all – including people that the logistician would rather have informed in a more considered way, preferably once the problem had been solved!

Two-way radios also have the disadvantage of being battery-operated. The batteries need to charge overnight, and there will need to be spares available in the day for heavy usage. Another problem is that these radios are notoriously hard to control and audit. When a radio user misplaces their radio, they will casually pick up a different one. Who lost what can be impossible to determine by the end of the day. Besides this, they can seem attractive to certain nefarious characters and are often stolen. Good housekeeping requires that radios must be numbered and signed out. A lockable filing cabinet (with a suitable power lead access so that they can be charged) should be used to store any radios not issued out. It must also be remembered that a licence is needed to operate radios legally.

The advent of the mobile phone, which is lighter than a radio and has a longer battery life, has done a huge amount for communications on site, and they are now an absolute must for construction and logistics managers. The rising capacity of mobile technology has made the transmission of numbers, pictures and other information ever easier. The construction industry were early adopters of personal digital assistants (popularly known as 'palm pilots'), and now use the ever-more sophisticated offerings, which come with such functions as satellite navigation. Mobile phones should not be used in certain situations on site, for example when operating equipment or vehicles or when working at height.

The capacity of mobile phones to be used as cameras has many advantages for the logistician. For instance, if a delivery looks damaged, it will be photographed on a mobile. The picture can be transmitted instantly to the trade contractor or client, allowing the logistics team to determine whether they should receive the goods, turn them away or wait for the contractor to come and examine them. This cuts down on disputes further down the line.

Emails are also proactively used in construction as a way of minimising disputes. This means that they sometimes don't really need to be read, except in cases of a problem, since they are primarily sent out to act as an audit trail or to confirm a verbal agreement. In this age of information overload, it's important to be able to pick out what is genuinely urgent within an overflowing inbox, and trying to read everything can be impractical. Conversely, the person sending an email cannot assume it has been read or understood.

It's the author's firm belief that there is no substitute for talking face to face, but it's impossible to deny the advantages of technology for remote communications.

Still, as with all forms of communication, both the person talking and the person listening have a responsibility to ensure that shared understanding ensues. In the fast-moving world of construction, it's all too easy to take something at face value, when both misinformation and misunderstandings might be occurring. Time spent ensuring clarification is never wasted.

Signage

Signage may not sound like the most exciting subject in the world, which would explain why so many projects are peppered with ad hoc signs! In fact, signage should be a vital strand in your communications strategy.

This strategy should be planned out carefully, rather than allowing the majority of signage to be created reactively once a need is identified. Signage, after all, plays two very important roles. First, it helps keep people safe. Second, it improves efficiency by providing information and directions.

Of course, some signage is mandatory in any case. Signs showing fire exits, 'no smoking' signs and other safety information need to be displayed by law. There is a huge variety of signs that may be required, some of which will probably be purchased from specialist sign manufacturers. They will have extensive product catalogues of all standard signs.

However, some signs will need to be bespoke. There are two ways of producing these. If a substantial sign is required, for instance because it is going to be external to the site or on its perimeter, it is probably advisable to get it made up by a specialist manufacturer. However, for internal signage, it's also possible to buy specialist software packages, which allow you to create your own signs. These can then be printed and laminated, for use on walls or hoardings. Signs can be very sophisticated – they can be illuminated or glowing – or basic printed signs with a bit of weatherproofing.

Since external signs are most likely to need specialist manufacture, it's sensible to consider these first. Generally, signs will get smaller as you enter the site itself, because people are physically closer to them.

It should be remembered that people can be blinded to signs if they're bombarded with them, so signs do need to be visually stimulating. Some will have to be duplicated to reflect the languages spoken on site. Commonly, it is not possible to have signs in all workers' mother tongues, but there are several common languages – for instance Urdu, Russian, Spanish and Chinese – that are widely understood by many nationalities. When building in Wales, all signs must be reproduced in Welsh – whether or not there are any Welsh-speakers on site. And, of course, pictorial signs overcome most language barriers, as shown in Figure 6.3.

Alongside the need to ensure that everything is reproduced in the correct languages, it's important to think closely about the wording, to make sure that information is conveyed as clearly and succinctly as possible. It's very easy to take a piece of information for granted, when it actually needs to be made explicit to the reader. It's often a good idea to ask for a second opinion, just to check a sign says what you hoped it would, before you get it made up!

Some signage relates to general communication with staff, so that they are aware of accident statistics, training courses and other helpful information. Other signs have a very terse – and possibly life-saving – message for those on site.

Of course, clarity of communication is particularly important when the sign relates to health and safety. This will include signs warning of hazards, such as holes and falling objects. Many other hazards – electrical and chemical – will need signage, too.

Other mandatory signs relating to health and safety include those advising of the need to wear hard hats and other personal protective equipment (PPE). Some signs are mandatory by law; others are mandatory in the sense of indicating what workers MUST do on site. Prohibitive signs, on the other hand, are there to tell staff what they must NOT do – no smoking, no radios, no parking, etc.

Parking signs are not, of course, the only necessary signage relating to vehicles. Movement of both vehicles and people in a safe and efficient manner need to be managed by a carefully considered approach to directional signage. Wherever possible, road signage should exactly mirror that which would be used on the public highway.

Figure 6.3 12 Essential Rules' poster created for the 'One in a Million' safety scheme is shown in Spanish, but was also available in English, French, Russian and Urdu

Directional signage

Some non-statutory signs are nevertheless vital to the efficient running of the site. Directional signage, particularly on large projects, has a vital contribution to make to overall productivity.

In construction, floors are referred to as 'levels' and areas are divided into 'zones', which may be named by colour or, perhaps, by something more imaginative should the project manager feel thus inspired. Both the signage and zone names should be informed by one key fact, however: most people on site will be strangers, most of the time. Indeed, sites change so dramatically, so quickly, that even a regular worker is almost always a stranger to the majority of the site, except at his or her workface.

The complexities of construction mean that routes that exist one day will have moved the next. It is not uncommon for workers to spend a sizable chunk of time trying to find their way back from the canteen! It's not that surprising when you consider that new routes are forged to accommodate the shifting uses of space and structures during the building process. However, the worker in question is still likely to feel that they should be able to find their way back to their area of work unaided, and is most unlikely that they will ask for directions.

This tends to be an accepted part of site life. Remembering the sheer size and scale of large projects – Terminal 5 at Heathrow Airport was built on a site the equivalent size of 54 football pitches, for instance – getting lost seems almost inevitable. Such sites are dynamic, vast movable feasts of corridors and alleys, both above and below ground.

Walls, corridors, doors, accommodation and temporary offices will all move many times in the course of such a project. Even on the smallest sites, things can change suddenly, so that what was once a thoroughfare becomes a blind alley. Similarly, even projects which appear very open, such as a modern shopping mall, depend on a warren of corridors behind the scenes.

This is why many signs, although non-mandatory, are still vital to both the safety and the efficiency of the site. Obviously, as the construction develops, the signs need to change accordingly, which means that supervising and updating the signage is an ongoing logistical task.

Directional signage will probably be needed beyond the perimeter of the site itself. The logistician must ensure that there is adequate information beyond the site for delivery drivers, especially if the drivers come via a particular route agreed with the Local Authority. Such signs will often need to be a considerable distance from the site itself, perhaps some miles away. It is common to post directions coming off the local motorway. Signs alerting drivers to the fact that a route is for construction traffic only may be a necessary part of the signage strategy.

Of course, the logistician must be prepared for the fact that signs telling other drivers not to enter may well be widely ignored. Signs saying 'don't climb' or 'don't smoke' can be similarly ineffectual as a way of influencing behaviour.

Nonetheless, the logistician will have successfully fulfilled their role by displaying a comprehensive suite of mandatory, informational and directional signs that constitute the backbone of the day-to-day risk management of the site.

References

Emmitt S and Gorse C (2003) *Construction Communication.* Wiley-Blackwell, Oxford.

Higgins G and Jessop N (1965) *Communications in the Construction Industry: The report of a pilot study.* Tavistock Institute, London.

Moore GE (1965) Cramming more components onto integrated circuits. *Electronics* 38(8): 114–17.

Chapter 7
Managing Critical Risks

Good logistics undoubtedly has a positive impact on productivity, but it should also be remembered that it can bring huge advantages to a project through a streamlined and holistic approach to some of the most important issues on site. This chapter explains how the site logistics team can help create a coherent service to manage fire risks and first aid. There is also an explanation of how good occupational health services can benefit both the health of the workforce and project performance.

Importance of fire management

Managing the risks associated with fire is one of the most important parts of the site logistician's remit. As with all aspects of the logistics, this requires a well-thought-out strategy and a thorough risk assessment if it is to be done well.

Causes of fire

Attitudes to fire management in the UK changed forever in the wake of the King's Cross, London fire of 1987, where 31 people lost their lives and many more were injured. The cause of the fire was believed to be a discarded match on an escalator. Smoking on the escalators into Underground stations was banned thereafter, and started to be seen as far less acceptable on construction sites, but the scale of the tragedy also led to far more attention being paid to all kinds of fire risks. Deep-fat fryers ceased to be a mainstay of site canteens, and unprecedented research attention was focused on issues such as the development of fire-retardant temporary accommodation.

It has often been stated that there are three main causes of fire: men, women and children. However, this light-hearted aphorism conceals a serious truth. Human behaviour has a major bearing on fire risks, and the compliance of the workforce with 'no smoking' rules and the regulations safeguarding high-risk activities, such as hot works, will be crucial to the avoidance of fire. At worst, human behaviour can extend to the deliberate setting of fires. Arson attacks have been attributed to both rogue security firms and animal rights protesters (Monaghan 2006; Anon 2008), among others. Arson can have a purpose such as intimidation or protest, but it can also be caused by youth vandalism. Very young arsonists – primary school-aged children – may not even be acting with any malicious intent. They are simply at an age where they are fascinated by fire, and only dimly aware of the potential consequences of their actions.

Besides deliberate intent, there are many other causes of fires and many other factors which can increase the risk factor of a blaze. Fire risk management is one of the primary arguments for the efficient and timely management of waste disposal on site, to prevent a build-up of flammable materials, and should be a key part of the fire prevention strategy. Mastic glues and wood, to name but two examples, are highly flammable materials, and common sense dictates that unnecessary quantities should not be left around a site.

Waste also includes packaging, much of which is also very flammable, and this should be constantly monitored. In addition to packaging, protective materials used to cover finished surfaces in areas still under construction can be disastrously efficient as kindling. If non-fire-retardant covers are not used, they can become a major fuel source. In 1991, a major conflagration at Minster Court, London, where the plywood protecting the escalators caught light, very nearly led to the loss of life and caused damage reportedly worth £105 million pounds (Puybaraud *et al.* 1999). This prestigious project was nearing completion so the fire also caused significant delays to handover.

Welding, cutting and grinding are also high-risk activities. Such activities will, in the UK, be covered by a mandatory hot works permit system, which applies to any work normally involving sparks or naked flames. It should only be authorised when there is no safer way of working, and must be carried out by competent personnel and with appropriate fire extinguishing facilities on hand.

Managers should be aware that some tools, such as petrol-driven angle-grinders, which should notionally not be brought onto site, may appear within the personal toolkits of workers who have previously been used to working in environments where such items are still in common use.

Whilst many site activities are hazardous, the logistician should not overlook fire hazards relating to welfare areas and accommodation units, where ovens and other sources of heat will be present.

Finally, the logistics manager needs to be particularly aware of the dangers posed by gas bottles and other highly flammable materials, such as chemicals, mastics and paints. There needs to be a robust and recognised system for storing gas bottles, with additional fire prevention and suppression measures in place. Wherever there are high-risk materials stored, which should be outside of the building, there needs to be as many prevention measures as are reasonably practicable. Indeed, gas cylinders should be stored three or four metres from any buildings – depending on the size of the store – and 1.5 metres from any boundaries. Valves should be kept closed, even on empty bottles, to prevent any residue from escaping, and sources of ignition, such as lights, should be at least two metres away. Appropriate signage is also vital.

The fire safety plan

The Construction (Design and Management) Regulations 2007 (CDM Regulations) stipulate that provision should be made for emergencies such as fires when planning for the construction phase. The construction phase plan should identify all major hazards, including fire risks, which the principal contractor will need to control during this phase of the project (Furness and Muckett 2007). Therefore, a fire safety plan is both a practical and legal necessity.

The key to a successful and comprehensive fire safety plan is to understand the building and the processes that will create it. What will it look like and be made of? What will be happening to it, and what materials will be involved? What type of packaging will these materials be transported in?

Due to the radical transformation that a site undergoes during the life of a project, the fire safety plan needs to be constantly updated. As noted in Chapter 6's section on signage, it is easy to get lost in the ever-shifting temporary corridors of a site. This is inconvenient and bad for productivity under ordinary circumstances, but in the case of a fire the consequences could be tragic. Not only does the workforce need to know how to get out: the fire brigade needs to know how to get in. Therefore, time and effort has to be continually given to ensuring that the plan is relevant and up to date.

Unfortunately, since a fire plan, by its very nature, is not in regular active use, there is a danger that other 'more urgent' tasks can take precedence. However, it must be constantly remembered that there will be little warning in the event of a blaze, and certainly no time to update crucial safety and rescue plans. Logisticians should therefore pay due respect to this task, and have the best fire plan possible in place at all times.

The fire safety plan and its regular updates should be issued to all contractors coming onto site, and explained in all routine safety inductions.

The fire safety plan must be supported by comprehensive signage, fire points, fire-fighting equipment, exit routes and muster points, which should be external to the site.

Adequate fire extinguishers also need to be provided. There is an eight-part series of British Standards, BS 5306, that covers fire extinguishing installations and equipment. Foam is suitable for many types of blaze, and can extinguish a fire by smothering it, but CO_2 (carbon dioxide) is the only type of extinguisher that is suitable for electrical fires. Faced with a fire, even the trained person can find if difficult to know what to do first, whether to raise the alarm or fight the fire. What extinguisher should they use and how do they operate it? A programme of training and drills should help to resolve this issue, but should be made as realistic as possible.

A standard fire point will consist of two 9-litre foam extinguishers and one 2-kilo CO_2 extinguisher. These should be mounted on a bespoke frame stand, which would generally be painted red. There should also be a space for signage and a copy of the evacuation plan, as well as a rotary bell or other form of fire alarm (Figure 7.1).

Figure 7.1 A standard fire point from Bow Bells Project is being audited

A question frequently asked is how many fire extinguishers are required on site. This is largely about common sense, and every site is different. As a rule of thumb, one fire point should be provided for every 250 square feet in open-plan areas, but of course where there are corridors and stair cores you may need more. This is a judgement call, which you should make with your health and safety adviser. Construction sites are not always about building: sometimes they're a hole in the ground. However, the site will still need to have fire points that are easily accessible and identifiable by the workforce. These could be placed around the perimeter or by the gates.

As well as extinguishers and fire points, the logistician must ensure that there are adequate fire mains. Dry risers, which rely on the ability of the fire service's specialist equipment to draw up water, are only sufficient to a height of around 50 metres (Furness and Muckett 2007), and so it is likely that wet risers, which are constantly charged with water, will be required on certain sites.

It is also important to maintain productive liaison activities with the local fire brigade and the site's neighbours. The logistician needs to ensure that the site's muster point is not inadvertently designated as an area already being used by neighbours, especially considering that any fire could quite easily affect both the site and its neighbours. Even without this consideration, liaison should in any case be part of the good neighbour policy.

Evacuation plans

The most important element of any fire safety plan is working out how to get people out of the building in a safe and timely manner. Ultimately, responsibility tends to sit with the construction manager, but on larger sites the logistics manager is quite commonly in charge of this function. This may then be delegated to a member of the supervisory team, so long as everyone involved in the fire response is adequately trained.

In a perfect world, when the fire alarm is raised, people should exit in an orderly fashion, and gather at the designated points. In reality, many people tend to assume it's a drill, and stay to finish off whatever they're doing and collect their things.

Fire travels incredibly fast, however. Footage of the fire at Bradford football ground showed that the building was engulfed in less than six minutes. It doesn't take long for a fire to reach the intense heat necessary to trigger a flashover. Many materials become hazardous when alight and can accelerate the spread of a fire. In modern buildings, speedy evacuation is also particularly desirable due to the sealed nature of the environment. If the ventilation system fails, people can suffocate extremely quickly. Therefore, making sure that all staff are aware of the correct way to react in the case of an alarm is vital. More people die of smoke inhalation than burns.

It should also be remembered that practical constraints might preclude the use of a standard, main-powered alarm. Where there is no electricity, a fire bell can still be effective, although it sometimes seems that vocal cues – the sound of someone shouting 'there's a fire!' – actually trigger the swiftest response of all! However, best practice is likely to involve procuring a wire-free alarm, and these are now becoming more common. Unlike the old temporary wired-alarm systems, they cannot accidentally get their power source cut.

The logistician may also want to consider instituting a two-stage alarm. This can ensure that if there is a break glass or smoke detector activation that wasn't actually triggered by a fire the whole site is not evacuated unnecessarily. In a two-stage system, the alarm is only sounded in a permanently staffed security office. Security can then radio the fire marshal for the zone that initiated the alarm and will then immediately sound the full alarm if a fire is confirmed.

Another issue around evacuation that the logistician needs to be aware of is the difficulty of establishing exactly who is on site at any given time. A number of construction site access and control systems can be used to verify who is on site on any given day. The problem with that type of system is that there will typically be many ways on and off a site, both formal and informal. Temporary fencing and loading bays are just two examples. Workers can have a variety of reasons for wishing to circumvent the system, from losing their pass to a temperamental disinclination to cooperate with anything pertaining to management information-gathering. From the logistician's perspective, however, their motives are less important than the practical implications – in short, a printout of who is on site can be useful, but it's most unlikely to be definitive.

Therefore, it's not enough to rely on access control systems; the best endeavour of fire wardens and marshals is also required to ensure that areas are actually cleared. There is a responsibility on supervisors for each individual contractor to see that their workers are accounted for, but on a large site it's still very difficult to know exactly where everyone is. This is not helped by the fact that in the event of an evacuation many workers may be tempted to go straight home. Awareness-raising of how grave the consequences could be for members of the emergency services who might attempt to rescue them if they are unaccounted for should form part of the induction, but cannot be relied upon to create blanket buy-in to best practice.

Where formal entrances with turnstile access to the site are concerned, the security staff have to be aware of the need to switch them to free access as soon as an evacuation begins. Gates should be opened to prevent people having to navigate access control measures.

Fire exits should be reviewed to ensure that they don't lead to dangerous areas. This may sound obvious, but routes straight into main roads are not unknown. The author has previously had to advise of potential improvement strategies in the fire plan on sites where exits were leading to railway lines and sheer drops. Also, however tempting and convenient it may seem, fire doors should NOT be propped open with fire extinguishers. It's dangerous and bad practice, and the logistics manager must strive to create a culture of awareness where such common but inadvisable practices are not tolerated.

Fire marshals and fire wardens

The fire marshal has a range of duties, and it's important to ensure that the responsible person receives adequate training for the job. Therefore, they should have completed a certified course. They need both to understand their own role and to have some knowledge of the behaviour of fires. Limited fire-fighting techniques will also be taught.

The fire marshal, in conjunction with the construction manager, will be responsible for the creation and maintenance of the fire safety plan, and will be in overall charge of the fire wardens. They should test the fire alarm every day, and carry out periodic evacuation drills.

They need to make regular checks to ensure that all extinguishers at the fire points are fully charged and in good working order, and that signage showing evacuation routes and fire exits is all clear and up to date. It is likely that these will need to be updated on a daily basis. Crucially, their plans for evacuation routes and exits must also be updated daily and stored outside the building within the overall safety plan. This is so that the fire service crew can find their way into and around the building in an emergency. After all, construction sites can be hazardous at the best of times. In the event of a fire, if the emergency services need to enter the building they will be in an unfamiliar area and are likely to have significantly reduced visibility. They will have enough to contend with without the added complications that could be created by blind alleys or unexpected shafts. It is good practice to liaise with the local fire service and allow them opportunities to familiarise themselves with the project at certain stages of progress. The fire service should know where keys are kept for stores and gates etc., especially if night-time or weekend security staff are not present. Particular good practice would allow the fire service to carry out mock rescues as part of their training and understanding of the physical constraints of a site. The contractor should also ensure at all times that materials, plant or equipment are not impeding access for fire service vehicles – especially adjacent to designated 'crash gates' within the perimeter fence or hoarding.

The fire marshal controls and administers any hot works permits. Monitoring hot works with regular patrols is a crucial part of the job. When the work is completed, the fire marshal will be the one to check and sign off the hot works area. Similarly, they need to monitor the use of flammable materials, particularly gas, to ensure that bottles are safely and properly stored.

Fire wardens should also be fully versed in the nature of their role and its responsibilities, and therefore need adequate training. It's particularly important to make sure that fire wardens understand that, whilst they're there to assist, they, like the fire marshal, must never put themselves in danger. They will also need regular briefings on any changes to the fire safety plan.

However, whilst the fire marshal has an active fire-prevention role, and will spend a significant portion of time patrolling and checking the site, fire wardens will have other jobs to do. They are there to react to the best of their ability to an emergency situation, using their best endeavours to assist in the evacuation prior to the arrival of the emergency services.

Contractual barriers to a seamless response

Construction has a culture of dispersing risk down its supply chain, and this can create particular challenges.

From a fire safety point of view, this can be particularly problematic when it comes to the provision of specialist equipment. Whilst the principal contractor will be responsible for the overall site, trade contractors are often expected to manage the risks of their own specialisation – such as hot works.

This dilutes the ability for any one point in the chain of command to ensure that every risk is covered. If a trade contractor neglects to bring their own equipment with them, the potential loss of productivity that could arise from this oversight may tempt them to either work without it or borrow equipment from one of the main fire points. This equipment will frequently have to be discharged in the course of hot works for the purposes of cooling or containing a localised fire. Theoretically, they should then replace this equipment. In reality, if there is a further oversight, this creates a gap in provision, and an extra risk that will require increased vigilance on the part of the fire marshal to manage. Ultimately, the principal contractor may well end up funding replacement equipment.

Clearly, this is not the optimal approach, which is why it is becoming less common on larger sites. On the average site, however, it's still all too common. These costs may appear relatively trivial at first sight, but in reality the costs of signage, extinguishers, personnel and other fire safety provisions can easily be a five-figure sum.

Therefore, it is argued, a holistic approach is far more desirable. If there is no attempt to pass on these responsibilities, the logistics team can create a mobile fire point, which can be wheeled to the site of any hot works or other high-risk activities. This can be signed out whenever necessary and the trade contractor can be routinely billed for any equipment they have used. Rather than informal borrowing of site equipment, there would instead be a procedure and an audit trail. It's more efficient, and it's far less risky.

Introduction to first aid and occupational health

On larger sites, the provision of equipment and the first-aiders will often fall under the remit of the logistics function, or a junior manager may otherwise be given responsibility for this. This section provides some pointers on what should be considered by any member of staff given responsibility for organising the first-aid response. It may sometimes be delegated to junior managers, but it is nevertheless one of the most vital resources to be procured for a site.

Creating the right level of first-aid provision

The statutory minimum provision for first aid is to have one first-aider for every fifty employees. If there are fewer than fifty, an appointed person is deemed sufficient to take charge of the first-aid box, which must be provided. In the view of the author, however, compliance with the regulations is unlikely to provide the best possible response to every contingency that might arise on site. Legal compliance, therefore, will keep company directors out of jail, but it shouldn't automatically be considered enough to make a firm into a caring employer. There's a substantial gap between the legal requirements and the provisions put in place by organisations aiming to be among the best in this area.

The good news in today's industry, however, is that as companies become more socially aware, and as health and safety climbs further up the agenda, appropriate first-aid facilities are put in place on an ever-increasing percentage of construction sites. Actual provision levels, assuming that they exceed the minimum requirements, will be largely dictated by the company policy of the principal contractor, and by the project's needs in the view of the site team.

In order to provide comprehensive care to all staff, it's best to think less about compliance with current requirements and more about a holistic approach to the occupational health of the workforce.

Whilst exceeding minimum requirements is advocated, the actual level of provision will still necessarily be informed by the size of the workforce. If 250 workers are on site, it may well make sense to provide a dedicated nurse to the site, whereas much smaller sites can reasonably be resourced with first-aiders.

The logistician therefore needs to begin by working out how many people are going to be on site at any given time, since this will fluctuate over the life of the project. By consulting the labour histogram, the likely peaks in staffing levels can be identified and appropriately resourced. It may well be that minimal resourcing will be appropriate at the outset, but plans must be in place to upgrade the provision of first-aiders, and the provision of first-aid equipment, as the site gets busier. Besides this, the logistician should take account of the type of works being planned, and their likely hazard levels, when planning resourcing levels.

Another important factor to consider is how the team will respond to those working in isolated parts of the site, paying particular attention to any lone workers. Shift work is another aspect to consider – it's all too common to see state-of-the-art facilities Monday to Friday but inadequate cover for the nightshift or at weekends. As an industry sector, we don't always replicate services out of hours to the extent that we should. The logistician should be aware of the need to ensure that adequate cover remains in place if a first-aider takes a holiday or is absent on sick leave. One of the ways of covering absence, sickness and shift work is to have a portion of the security staff trained as first-aiders. The author works on the principle of aiming to have 50% logistics and security team on any given project trained in first aid, as these are the people who are nearly always present on site.

First-aiders need adequate training. Indeed, the provision of trained and suitable people is a legal requirement. The Health and Safety Executive (HSE) approves training companies and other training providers. Therefore, whether training is delivered by an organisation such as the Red Cross or by a commercial provider, there is a system in place to ensure that it will be of a consistent standard. As long as the training provider is HSE-approved, there should be minimal variation in quality. There are two relevant courses. The course for an appointed person, who might be responsible for first aid on a very small site, takes one day. The full first aid at work training, which should be used for all first-aiders, takes four days.

The logistician should ensure that the training received is applicable to the nature or the hazards of the workplace. It must also be remembered that training and qualifications in first aid have a lifespan, and that first-aiders need to attend refresher courses regularly. If the refresher course is not undertaken in time, the first-aider will have to take the entire course again.

As always, the issue of who is going to provide these resources is a question which must be settled. The decision will be affected by the contractual allocation of responsibility. The principal contractor will be expected to provide a certain level of cover, but the subcontractors will supplement this with their own first-aiders and specialist equipment. After all, they have a statutory duty of care to their staff. Nevertheless, as specialist logisticians become more common on construction projects, they are increasingly likely to be handed responsibility for ensuring that the overarching site provisions are put in place. They and their staff will be likely to lead any emergency response, since, as we shall see, logistical issues are often amongst the most challenging – and the most vital to resolve – when dealing with any serious event.

The logistics manager has to make sure that it's easy to identify first-aiders. Ideally, they should have their pictures and mobile phone numbers displayed prominently on-site noticeboards. Green crosses on hard hats can also be helpful, while some sites use high-visibility vests with 'first-aider' emblazoned across them. However they're marked out, it must be remembered that it is crucial to be able to locate them quickly. The person responsible for coordinating first aid and incident response, in particular, has to know who they are, regardless of which company they work for, and how to get hold of them. It makes sense to ask the first-aiders from all organisations to work as part of a wider team, and to make sure that there's an informal agreement that all first-aiders are willing to respond to all incidents on a project.

First-aid equipment

First-aid equipment and facilities can range from a small box to a fully equipped medical room; once again, it depends on the size of the site and the degree of difficulty of the job. Proprietary equipment can be bought off the shelf (there are regulations around what must be included and excluded from a first-aid box), whereas some sites may have a comprehensive kit of rescue equipment. Some common remedies are now banned from first-aid boxes: a first-aider does not have the necessary training to administer a paracetamol, whilst some of the more specialist equipment needs specialist training, to ensure that it doesn't cause more harm than good. Aspirators would be a good example of this type of equipment. When the project demands a full-scale emergency room, the nurse's grade and/or experience will determine exactly what can be stocked. Not all nurses are qualified to do everything – sounds obvious when stated, but it can be overlooked when resourcing decisions are made. In terms of larger equipment, it's particularly important to obtain the right kind of stretcher. On a construction site, a stretcher that can be lifted and carried at various angles is essential. Local ambulance stations can be approached for advice on exactly which type should be purchased.

First-aid boxes are available off the shelf in various sizes, depending on the size of the workforce they will be needed to cover. Recently, travelling first-aid kits that fit onto belts are becoming more popular, since they're both easily accessible and secure. It's an unfortunate fact of life that, as a society, we're not always as respectful as we should be of this type of equipment. Elements of first-aid kits can easily go missing, either because they've been used and not reported or because they've simply been taken. Therefore, supplementary equipment as a back-up needs to be part of the first-aid plan, and equipment should be regularly audited. Similarly, first-aid rooms should have a regular maintenance and cleaning programme, as there needs to be the right clinical environment. It's also important to keep equipment sterile, and to remember that some equipment will have particular storage requirements and a limited shelf life. Regularly check for expiry dates where relevant.

Thought needs to be given to the positioning of first-aid points. If they're miles from anywhere, they're not much use. A major site might easily need six or seven satellite facilities. Mobile first-aid points can also be an appropriate way of creating cover for large areas.

Dealing with emergencies

Responding to an emergency on a construction site can be far from easy. Wherever people are working, the person in overall charge of the response – often the logistician – needs to be asking themselves how easily first-aiders could respond to an incident in that environment. How will they gain access? How could a stretcher be brought in if it were needed?

As with many aspects of logistics, this requires a high level of awareness about what's going on around you. If there are a significant number of people on site, the balance of probabilities suggests that, despite all best efforts, you may be required to deal with an accident or serious illness at some stage.

In the case of major incidents, it's important to reflect on the impact this might have on the local casualty unit. This requires third-party liaison, so that the relevant parties understand what you might need, and you understand their capacity.

It's important to be aware of the limitations of paramedics and firefighters, who have the right to refuse to put themselves in a dangerous situation. This can sometimes take precedence over treatment. Paramedics have refused to attend scenes which need to be accessed by ladder. This is a potentially very serious situation, and it's the ultimate responsibility of the project director, rather than a first-aider, to find a solution. Either the paramedics have to be persuaded that they can reach the scene safely or the manager in charge needs to find a way of getting the victim to an accessible area. There is no one-size-fits-all answer to such difficult and dangerous situations, but a solution will certainly have to be found. This takes strong leadership skills, and an ability to think clearly and calmly in a crisis.

Crisis situations take their toll on all involved, and this is something to be borne in mind when selecting first-aiders. It's not a good idea to just volunteer the nearest person for such a role. They need the right temperament to cope under pressure. Being trained in first aid does not necessarily mean that someone will be able to put that training to good use: some may be incapable of giving first aid when confronted by the visceral unpleasantness of a major accident scene.

Thankfully, the industry has made huge leaps forward in terms of reducing its accident rate. The author can remember when it was common to have an ambulance attend site about once a month. Now, such visits are far rarer. The one drawback to this achievement is that as major incidents decrease the response team gets less practised at dealing with them.

One way to overcome this is by using scenario training. Running regular exercises or rehearsals is strongly recommended. There are companies who will come on site and create an accident scenario using stage make-up. It's not quite the same as real life, but it is still a useful and worthwhile exercise. There is a body called the Casualty Union, whose members are very well practised at pretending to be injured. Nothing could quite recreate the smell and panic of a really bad accident, but at least the team can get used to practising their skills on people who appear to be in real distress (Figure 7.2).

Strength and agility can be a useful characteristic in first-aiders. The author once saw a site nurse climb a 90-metre tower crane to assist in the rescue of a worker suffering from a back spasm. The worker in question was hardly able to move, and another tower crane had to be used to get him into a rescue cage.

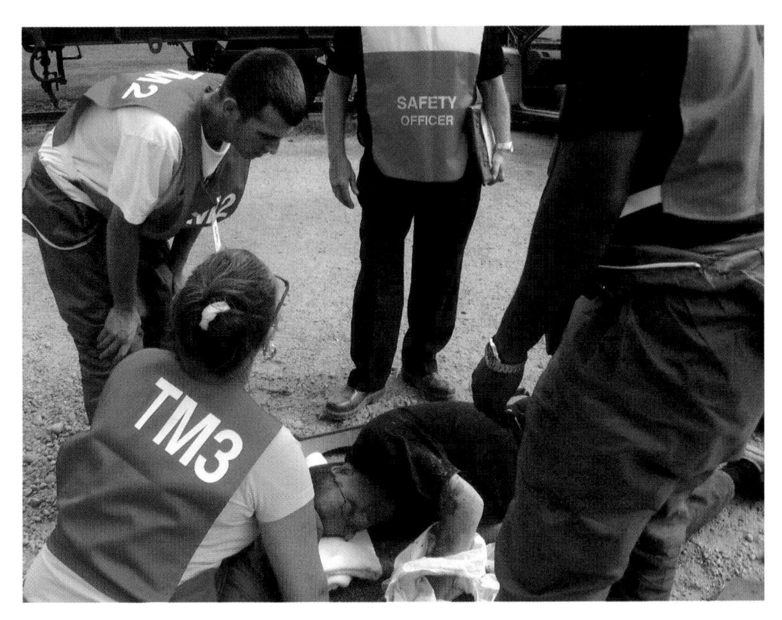

Figure 7.2 Staff are faced with a casualty in an emergency response training session held during a company-wide health and safety competition

One more point to bear in mind when selecting the first-aid team is that it's good practice to have a balance of men and women on the team. Both men and women can be casualties, and there may be religious or cultural sensibilities about being treated by a member of the opposite sex.

Finally, remember that first-aiders are people who have attended a four-day course. They are not always trained in incident management, or in recovery and rescue. The duty of a first-aider is to do all they can to keep an injured person alive until professional help arrives. They are not paramedics, and expectations of their capabilities can be unrealistically high.

Occupational health

As already stated, when there are a significant number of people on site, it's good practice to have a nurse or paramedic in attendance (Figure 7.3). The next step up from that, in terms of provision levels, would be to employ a nurse or medic qualified in occupational health.

Occupational health specialists hired for construction sites often tend to be nurses, but they will be able to tap into a network of other services by making referrals.

So, why would it be desirable to spend the extra money required to hire an occupational health specialist?

First, of course, it is an opportunity for employers to take the very best care of staff as possible. The service can be perceived as of real value

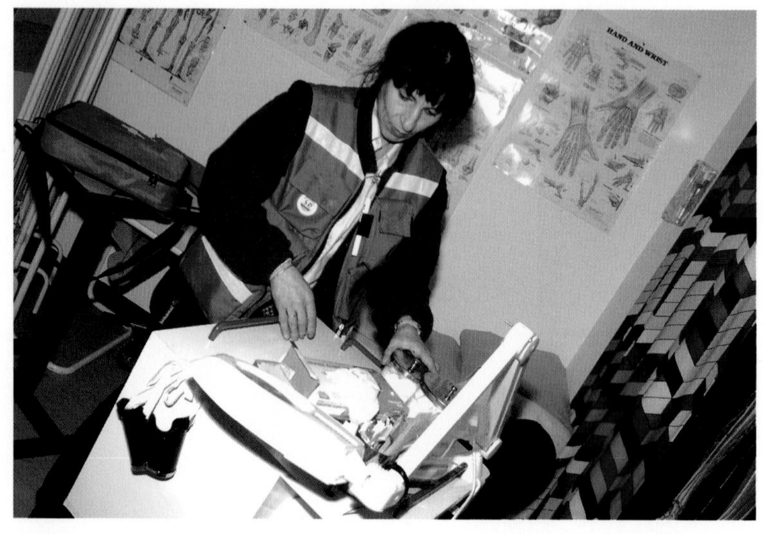

Figure 7.3 Unilever House Project site nurse, Janine Berns, checks her equipment and first-aid consumables

by potential workers. Therefore, in times of high employment, it could be a driver for selecting work on a particular project. It's also of value to existing staff, so it can help in reducing turnover, because they will feel more valued and respected by their employer. Our culture is becoming increasingly health-conscious, and occupational health, like better choices of healthy food in site canteens, has become more and more important to staff. This isn't just a perceived benefit – the last time the author was involved in such a scheme, three members of staff were able to gain early treatment after serious illnesses were detected.

Occupational health services can therefore be a beneficial part of any induction process. Questionnaires can be administered routinely, whilst key workers can be given a thorough medical. This means that workers can be screened for fitness prior to taking on key roles such as tower crane operators, where the operative needs to be fit enough to work at height. Eyesight checks are also advisable.

Another benefit from such a service is that it is likely to have an extremely positive effect on productivity and efficiency. Staff can get the advice they need in half an hour, instead of having to take half a day off to have a check-up. Many workers may be staying in temporary accommodation away from home, and may not even be registered with a local GP. If they're injured playing sport at the weekend, they may just take the day off unless they know help will be available. Occupational health service provision can encourage them to come in and get the problem resolved.

Lastly, providing appropriate occupational health facilities has been demonstrated to have the potential to save a substantial amount of money – although that shouldn't be the only motivation to do it. Still, there is a robust business case for this kind of provision, which may interest clients and construction managers.

There are measurable benefits, too. As well as reducing staff turnover, having an occupational health nurse on site reduces staff absenteeism. It has been calculated that 504 hours were saved on one typical city centre site, just from minor injuries on site being treated rapidly and effectively. Added to that, 566 hours that might have been lost while workers sought treatment for non-site-related injuries were also recovered. With other miscellaneous efficiencies added in, over a thousand hours were saved. The savings from that more than covered the cost of an occupational health nurse.

So compelling are both the benefits and the savings, in fact, that BAA's Terminal 5 project retained a sizable team of occupational health specialists, including a dedicated site doctor. The doctor even liaised with the designers in order to check for any particular ergonomic risks in the project.

It was reported by the healthcare provider in a case study for the HSE (Pugh 2004) that this 'significantly raised the awareness of OH issues in designers' minds'. This has led to greater emphasis being placed on occupational health risk elimination/reduction both in the production phase and operational phase of T5. The Occupational Health Service is also involved with construction production leaders and, indeed, procurers. The aim is to ensure that occupational health risks are eliminated or reduced as far as is reasonably possible. Even if few projects can replicate that scale of involvement, providing occupational health facilities can still have a huge benefit to both the workforce and the employer.

Procuring occupational health services may very well fall under the logistician's remit, but specialist advice regarding the correct service to provide is important. Where health is concerned, there can be ramifications, both legally and medically, if the service isn't totally appropriate to the needs of its users. Making sure the service operates confidentially is also vital, both for the sake of compliance and to maintain the trust and enthusiasm of the workforce. And it must be equally available to all staff, whether site- or office-based.

Construction organisations are likely to go to an external provider. It may be advisable, therefore, to ask a consultant to scope out the service as one piece of work, and then put that service out to tender. As the healthcare provider's client, however, thought will be needed to determine exactly what level of service is required. The best practice would be a well-person health check for all, but, whilst that would probably not be economically suicidal, many firms will not be able to commit that level of resourcing.

The nurse has to be someone who is confident talking about taboo and intimate issues with service users, and should have the capacity to devote time to improving the training of the first-aid team. Awareness campaigns, lifestyle checks and healthy-eating promotions are other initiatives which could be provided by this service. When you consider that injury treatment alone will probably fund the cost of the service, the range of add-value activities it can also provide starts to look particularly impressive.

References

Anon (2008) Who guards the guards?: Arson and intimidation in Glasgow. *Building*, 29th August, http://www.building.co.uk/story.asp?sectioncode=667&storycode=3121236, [accessed 10th January 2010].

Furness A and Muckett M (2007) *Introduction to Fire Safety Management*. Butterworth-Heinemann, Oxford.

Monaghan A (2006) Animal rights extremists threaten 2012 Olympic site. *Building*, 17th October, http://www.building.co.uk/story.asp?sectioncode=284&storycode=3075381, [accessed 14th January 2010].

Pugh C (2004) Occupational health provision in the construction sector. In: *Securing Health Together Case Studies*. Health and Safety Executive, Liverpool, http://www.hse.gov.uk/sh2/casestudies/healthywork2.pdf, [accessed 26th January 2010].

Puybaraud MC, Barham R and Hinks J (1999) Fire safety attitudes and management culture in the construction industry. In: Singh A, Hinze J and Coble RJ (eds), *Implementation of Safety and Health on Construction Sites: Proceedings of the Second International Conference of CIB Working Commission* W99, Honolulu, Hawaii, 24–27th March 1999. Taylor & Francis, Oxford, p. 134.

Further information

The Fire Protection Association (2006) *Fire Prevention on Construction Sites: The joint code of practice on the protection from fire of construction sites and buildings undergoing renovation*, 6th edn. FPA, Morton-in-Marsh.

Chapter 8
Security

Security doesn't always fall under the logistics remit. If there is a specific threat, it is less likely to be assigned to a general logistics contractor. Under normal circumstances, however, the shrewd logistics manager will want the security function under their wing so that they can maximise its contribution to operations on the site.

Traditionally, security has been something of a grudge purchase, like many aspects of logistics. Before the advent of the integrated logistics contractor, it would almost always be bought as a single service, and the award of the contract would be based on little other than price. At worst, this could lead to a security team on minimum wages being expected to be anti-terrorist experts, crime-stoppers and fearless fire-fighters. This was never realistic, and such a procurement strategy cannot achieve best value. Whilst, as we shall see, there are limitations to the duties a security officer can be expected to perform in a crisis, there is still a sense in which they are paid, not for what they do, but for what they may have to do in an emergency. When procuring security services, you've got to be clear about both what you want to achieve and what can realistically be achieved.

Remember that much of a security officer's work will be about providing customer service and being a first point of contact to all site staff and visitors. They will be the architect of all first impressions of your project. Therefore, you want staff who are well presented, articulate and confident. For this reason alone – although there are others – it's not wise to hire minimum wage staff for security. However many fine words are in the contract you agree with the service provider, the staff themselves will quickly realise they're paid somewhat less than the operative sweeping the site, and their morale and enthusiasm will be affected.

Remember also that a lot of available security jobs involve manning the reception of nice office blocks, where people are generally civil and there is usually effective heating and air conditioning. If someone has got the choice between standing in that office or at the entrance to an often cold, wet and miserable site, where people are often stressed and perhaps not always inclined to automatic compliance with rules and regulations, you will not attract decent staff without recognising the need to reward them. Therefore, principal contractors must consider whether it would be worth stipulating a minimum wage for the security team in the relevant tender documents.

The security industry has become increasingly regulated over recent years, and reputable security companies will be approved by the Security Industry Authority (SIA). Principal contractors should check for the correct accreditation compliance when procuring security specialists by visiting the SIA website (http://www.the-sia.org.uk/home/) and checking through its approved contractor scheme listings. Security companies need to be accredited by this authority. Bear in mind that individuals need to be approved too, and that different security activities require different licences. For instance, CCTV operators need a special qualification, and so do officers guarding licensed premises. Whilst this is relatively uncommon in construction, it could theoretically happen during the part refurbishment of major venues. There are also a number of International Standards (ISOs) and British Standards (such as BS 7499 and BS 7585) that a security company needs to comply with.

Construction sites don't lend themselves to a nice, swish reception area with a flawless filing system, although some of the best developments have proved that it can be achieved. Nevertheless, the correct approvals will need to be to hand for every officer involved in the security team. The SIA can carry out spot checks at any time, and all the relevant approvals will need to be in place and available for inspection.

Getting the right security team is important, as it is the first line of defence. The team will be the first response to all kinds of incidents, whether they are caused by criminals, terrorists, fire, flood or road accidents. Since such extreme situations only arise rarely, it's important to understand how they can contribute to the project under more normal circumstances. In order to create an effective security strategy, however, it's important to begin by understanding what the team can reasonably be expected to achieve – and what should not be expected of them.

The remit of the security team

Whilst security can be a grudge purchase, it can, when done well, create an indispensable resource for a project. Traditional security has been mainly about managing access (see below), meaning that the busiest part of the security officer's day has ended by 9.30 a.m. When the rest of the day is seemingly unproductive, as can happen with traditional security operations, it's understandable why it's seen as a disproportionately costly outlay, and as a barrier to production. Security is about managing

Figure 8.1 Secure logistics high-visibility jacket

risks by enforcing compliance with site rules, and, as such, it can sometimes seem frustratingly obstructive to other workers.

However, since the security team is going to be there anyway, and is in fact the most consistent presence on site, it's common sense to ensure that its members have as many skills as possible. A multi-skilled security team can provide fire wardens, traffic marshals, first-aiders, banksmen and hoist drivers (Figure 8.1). Suddenly, this team, equipped to deal with anything from accidents to late deliveries, is starting to seem like value for money. Still, it should be remembered that, although licensed security officers can be trained to perform other duties, banksmen and traffic marshals cannot perform security functions unless they've been licensed to do so.

Crime prevention is always a challenge on construction sites. Theft is not uncommon. The risks arising from this, like so many other aspects in construction, are spread through the supply chain. The principal contractor will usually provide a perimeter fence and gate security, but specialist contractors are responsible for protecting their tools and materials within that environment.

Generally, hoarding and fencing is a standard 2.4 metres or eight feet high. Plywood or temporary fencing may sometimes be sufficient as a barrier. In high-risk areas, other methods, such as anti-climb paint, passive infrared (PIR) lighting, CCTV or alarms may be added to increase security. Razor or barbed wire can only be used under certain circumstances. It is likely that you will need to consult with your Local Authority about what is and is not possible. It is a matter of project/client preference on the type of hoarding provided. Solid (timber) hoardings may require vision panels to be cut into them, allowing external patrols to see in. This has to be balanced against any desire for the site to be protected from the gaze of the public. Providing vision panels at various heights in the perimeter hoarding will allow genuinely interested and curious members of the public, including children, to safely observe the fascinations of the construction process. This facility can be enhanced by the contractor providing information posters and contact details about the project adjacent to the vision panels and will promote a positive image of the construction industry. The Considerate Constructors Scheme (and similar regional schemes) encourages contractors to create a positive image of the construction industry and provide comprehensive guidance through its code of practice (Considerate Constructors Scheme 2010) on how this can be achieved.

Unfortunately, wire or mesh fences provide little deterrent to the determined criminal, even when supplemented by regular security patrols. In the real world, creating a crime-free construction site is almost impossible. However, robust security measures will certainly reduce casual theft, and will make it as difficult as possible for crimes to be committed.

It must always be remembered that there are strict limits to what a security guard can reasonably be asked to do. There is a misperception that they are able to arrest suspects at will, but the law is, if anything, more rigorously applied to security officers than to the public at large. For this reason, whilst it's good practice to teach security officers conflict resolution and breakaway techniques (so that the officers can leave a situation safely if it escalates), it's very questionable whether training in more proactive approaches, such as restraint techniques, is necessary or advisable.

Even without such training, security officers can sometimes, for the best of reasons, take excessive risks in the discharge of their duty. The author employs a large of number of security officers, one of whom was once confronted by a visitor to the site office who pulled a gun. The security officer attempted to disarm the gunman, who luckily fled the scene, but such heroics could have very easily led to a fatality. Security officers have the same right to a safe workplace as anyone else, and the most that can be required of them in a conflict situation is that they call the police.

Clients still sometimes request that security officers be instructed to apprehend criminals such as teenage vandals. The police might once have been forgiving of such an intervention, but it would not be tolerated in the current climate. The security staff would be vulnerable to all kinds of charges, from assault of a minor to theft, should they confiscate a spray can. There could be a variety of most unpleasant allegations, and the staff involved might be prosecuted. Other jobs which people sometimes assume a security officer can perform are also off-limits. For instance, no one has the right to step out and stop traffic on Her Majesty's highway apart from police and Home Office-approved wardens. Clients, therefore, have to be educated in the limitations that govern the role.

Having said that, the current climate evolved partially in response to what were sometimes heavy-handed security interventions. This has led to the increased regulation of the industry. Security officers are required to have licences and background checks, and receive four days of basic training. This has to be a good thing in terms of creating a culture of professionalism in the security industry.

Another duty which can sometimes be performed by the security and logistics team is inductions. It is the principal contractor's responsibility to ensure that all persons entering the site have received a suitable safety briefing on site rules, and there are well-documented and established methods for achieving this. Even if the delivery of inductions is not delegated to the logistics or security team, they will probably be responsible for administering it. Inductions should be mandatory before the issue of a security pass.

The exact duties of a security team on any particular project should be very precisely defined. It's vital that the security team understand exactly what the expectations are, and that the specified tasks are achievable. For example, it may sound sensible to stipulate that all visitors should be escorted from the entrance to the site office. If the resources aren't there to achieve this, however, the rule will be ignored. Once one rule is disregarded, others can too easily follow, leading to the decay of process across the board.

Decay of process is the greatest threat to effective security. When there are rules, they have to be followed if the security operation is to be effective. Therefore, before the rules are put in place, there needs to be a sanity check: are the rules workable for everyone? If not, amend them. There's nothing more detrimental to morale and good discipline than rules being ignored.

Security officers are frequently put under huge pressure to bend the rules. If a manager has driven halfway round the M25 motorway to deal with a late-night emergency, they will expect to get onto site even if they have forgotten their pass. If they are denied access, they won't be very pleased. Sadly, some of the verbal assaults which take place against security officers are perpetrated by the staff who they've been paid to protect. Nonetheless, however crazy it may seem to deny access to a well-known member of staff, the officers will be trained not to make exceptions. For all they know, they could be dealing with a member of staff who was sacked the day before, and is now intent on sabotaging the site by way of revenge. The more exceptions they make, the more likely it becomes that one day the wrong person gets on site at the wrong time.

Of course, strict enforcement of the rules will sometimes have an impact on productivity – although far less impact than would be caused by a major security breach – which can be temporarily unpopular. This is stressful for the security staff. The logistics manager with responsibility for leading such a team must therefore make every effort to lead by example by carrying their pass at all times, and following all procedures to the letter. If there is any decline in standards, the logistician will need to gather the team together, point out the breaches and explain how important it is that rules are adhered to without exception.

It's good practice for the logistics manager to join staff on the gates when any new regulations are implemented, to ensure that everyone knows what's being done and why. They must take time to talk to their officers about any difficulties, and must come to their defence if they are harangued by other staff. It's also important to ensure from the beginning that measures are proportionate as well as workable, so that the goodwill of other workers can be maintained. Security has a close relationship to safety and it needs to be clear that the measures that are introduced are intended to benefit everyone on site.

This kind of proactive management is crucial to the maintenance of good morale, without which you can lose discipline alarmingly quickly. Once rules are circumvented, it's very difficult to recover the appropriate standards, so any decay of process has to be guarded against from the outset.

Managing access to site

Controlling access and egress are the core functions of site security, which makes the beginning of the day the busiest time for security staff. There are a number reasons why this is important. Not only is it done to prevent the wrong people gaining access in order to steal or cause other problems, it's also a key part of site safety. The principal contractor has a duty to prevent unauthorised people, specifically untrained people, from gaining access to a working site, where they could be injured. Remember, even if a burglar gets injured in the course of committing a crime, there could well be litigation arising from this. The law applying to children that get injured whilst trespassing on sites are even more onerous upon the contractor.

Therefore, the security team is the first line of defence of site safety, as well as site security. They will be in charge of checking that those coming onto site are suitably inducted, and that they have a CSCS (Construction Skills Certification Scheme) card (Figure 8.2). There are still some sites where this is not enforced, but there are now rules emanating from the Main Contractor Group that set out requirements for accreditation on their sites.

Figure 8.2 A security officer checks the ID of a contractor wishing to enter a project

Security officers are increasingly likely to be assisted by technology to manage access to site. Depending on the nature and size of the project, turnstiles are more and more frequently found on site. However, since low turnstiles can quite easily be hurdled, full-height turnstiles are now starting to be adopted. These can make access with tool bags difficult, a point which should be considered when selecting an access management strategy.

Also, when considering turnstiles, the logistician will have to look at the labour histogram and work out how long it will take to get each individual worker through the turnstile, past security and onto the site. You don't want queues round the block. Not only does that create disgruntled workers, it could also get in the way of passing traffic and pedestrian flow. Technology is starting to speed such systems up, with IT access control systems providing hands-free facilities, such as proximity readers. A proximity reader can approve a card without it even being taken out of a bag or wallet, although currently swipe cards are still in common use. It's also important to ensure that your selected technology has an anti-pass-back facility, so that groups of workers can't all gain access on one card. This is, again, driven by the need to ensure that all the people on site are bona fide and safety inducted.

When creating an automated system, the logistician needs to be aware of the requirements of the Data Protection Act. As soon as anyone starts to record personal details, they must be registered under the Act and comply with the rules regarding the using and sharing of data.

Construction is not yet always up to speed. One of the things principal contractors hope to gain from such systems is a record of arrival times and attendance, so that they can check who's on site, and ascertain whether their trade contractors are supplying the agreed level of staffing. However, this is where data protection comes into play. Before any data is shared, the right procedures and processes need to be put in place. There's a temptation to use the information for purposes that aren't appropriate. In fact, however, the client constitutes a third party and does not have right to any data without agreement from the data subject. Even the police, in fact, have no right to see such information without a warrant. Information can be distributed legally, so long as the uses of the data are made explicit and are understood and agreed to by the individuals concerned.

One of the new innovations currently coming into use is access management with biometrics data, such as fingerprints. There is a natural scepticism amongst many workers about providing such data, since they fear it could be given to others. That's one of the reasons why it's important to operate within the rules and use the legislation properly. You have to show the workforce that you are honourable and take your responsibility for their privacy seriously. CCTV is also covered by data protection, and individuals can request footage of themselves under the Freedom of Information Act. Don't think you can just fix up a couple of cameras as part of your security system. Without proper processes, you could easily be in breach of important legislation.

Used responsibly, technology can be of great assistance. It can add value and reduce costs. Technology itself cannot replace people, but technology and people in unison can be a very effective operational tool.

No review of access issues and security would be complete without the mention of keys. It often falls on the logistics manager, and their security team, to manage keys to access specific areas on site. A suitable system to issue and receive keys, particularly for areas where permits are needed, such as switch rooms etc. – will need to be put in place. Otherwise, if there is a lack of discipline, keys can get lost. They have to be signed out and back in. Otherwise, if decay of process sets in here, access to vital and perhaps dangerous areas will be compromised. It's also important for the protection of the security team, so it cannot be held responsible for missing keys or for irregular access. Because the team has access to all areas, it can be quickly blamed for issues that it had no involvement in.

There should be a particular access and security plan for a critical area that is, strangely, often overlooked – the site offices. Something in the nature of temporary buildings seems to make their users more casual about security, but there is usually a huge amount of intellectual property, mostly subject to commercial confidentiality, in the average site office. There's also usually a reasonable amount of desirable, expensive hardware around, including BlackBerries and laptops, which will possibly contain sensitive information.

Construction is not known for its clean desk policy. The nature of the business means that managerial staff are often distracted, or need to go out onto site, leaving behind high-value items in clear sight. Thousand of pounds' worth of equipment, plus the critical data they contain, can go missing.

Therefore, great care needs to be taken around offices, and there has to be a method for controlling access. The logistician should always be mindful of lone worker operations in offices, since this could render the worker vulnerable. CCTV is an option, but don't make the mistake of having the whole system in the same room – whole systems have been taken in the course of thefts in the past. Keep the equipment locked away so it can't be stolen or tampered with, and position cameras discreetly.

Night security

Night officers have a slightly different role from that of their colleagues on the dayshift. They will be the first line of defence when people attempt to enter the site out of hours. Therefore, their main duty will be to maintain patrols, and to practise awareness and observation. However, there are wider duties that a patrolling officer can perform, which can add value to the security function.

Pieces of plant may be operating throughout the night, or there might be a continued fire risk from hot works. A patrolling night officer should be instructed to monitor such issues as part of the site's standard operating procedures, so that problems can be discovered and tackled. The night patrol can watch for flooding potential, and be trained to switch off plant that's no longer being used. Conversely, the officer can ensure that plant that is supposed to be on is working consistently, providing basic servicing and refuelling for plant such as water pumps, so that their function is uninterrupted throughout the night. Even if the officer has to be paid more in recognition of the extra training and responsibility this entails, it will still work out cheaper overall.

On the night shift, it is sometimes very tempting to have just one guard in a hut monitoring several cameras. The author advises caution, however. There are definite risks to working alone, particularly in an out-of-hours situation.

Night security can also be managed by off-site support. It may be that, following an assessment, the logistician may decide that there isn't a need for a full-time security presence out of hours. One option may be a visiting patrol officer. Another alternative, which could be used alone or in conjunction with a visiting officer, would be a remote monitoring and communications centre. If there is an alarm activation or unusual activity on CCTV, they can call the emergency services or dispatch mobile officers to investigate. Sometimes, just turning up in van, and making as much of a fanfare as possible by switching on all the lights and instructing any trespassers to leave, is all the deterrent necessary. Indeed, lighting can be a very good deterrent to would-be criminals. The use of flood or PIR lighting would have to be by agreement with your neighbours. PIR lighting can be activated by birds and bats, so it is important to protect neighbours' privacy by ensuring that the angle of lighting does not shine onto their properties. Each site will have different circumstances to consider when arranging deterrents. The aim, however, is always to interrupt rather than to confront.

Even when there is an on-site security presence, the off-site team will still need to call in regularly. It is not unheard of for workers on unsociable shifts to nod off, but it's also important to make regular check calls to ensure that nothing untoward has occurred. To manage security officers' performance, it's possible to confirm patrol routes are being covered by having swipe card or barcode reader patrol monitors at various points on the route. This will confirm that all requested patrols are carried out. Officers should also call their monitoring centre as they leave for a patrol, and to confirm that they are safely back again. Technology cannot replace human contact for ensuring the safety of staff, however useful it is for performance management. If an officer fails to answer a call, immediate steps should be taken to ascertain that they are safe.

Dogs

Dogs are now being used more often as part of the security for certain sites. When sites are in a remote or difficult location, or in a high-crime area, it may be necessary to take additional security measures, one of which might be security officers patrolling with guard dogs.

There are many instances where dogs can be of benefit to the security strategy – as is well known, they are widely used by the police for detecting explosives and drugs. A security officer is unlikely to be using dogs for that, but it can still be extremely beneficial to have a trained dog to provide a local deterrent. Also, in the event of a major situation, it will be hoped that the dog will be able to protect the security officer.

Guard dogs do have to be specially trained, just like police dogs, so when employing an officer with a guard dog, it is important to make sure that all the right criteria are met. The officer needs to be a trained dog handler and the dog should be licensed in accordance with the Guard Dogs Act of 1975. It must also be remembered that the dog will need food, water and a suitable rest area, and that provision will have to be made to clean up after it. There will also need to be a comprehensive risk assessment and method statement.

Sometimes, it may be necessary to use dogs during the day, at special sites with particularly high risk levels, but they are more often deployed when the site is closed at night, largely as a deterrent to trespassers. However, just as with the security officers themselves, there are definite limits to what a guard dog can do. It is never appropriate to let the dog go. Even the police only do it when using dogs for searching. Similarly, an officer with a dog should never try to apprehend a suspect. All the officer should do in the case of an intruder is call for back-up, probably from the police. There are several reasons for this.

To begin with, it can be dangerous for all concerned, and handlers can easily get bitten by accident. Another issue is that if the dog bites a possible criminal, it would need to be demonstrated that this was appropriate and an act of self-defence. It should also be remembered that trespass is not a criminal act until another factor such as burglary is involved. Otherwise, it is only a civil offence. A trespasser can be asked to leave, or the police can be asked to assist, and they will ascertain whether there are any other grounds for arrest, but neither a security officer nor the police can arrest an intruder if trespass is their only offence.

Alarms

Dogs can deter intruders and support and protect their handlers. Besides this, they can act as excellent alarm systems. They have better hearing and sense of smell than we do, so they pick up the presence of intruders quicker than humans ever could.

Sometimes, more high-tech alarm systems will be installed. Alarms are very useful, particularly to protect offices and storage areas containing valuable items. Alarms can be simple, in that they can be either traditional sounders, or sounders and flashing lights, or they can be far more sophisticated. Alarms will alert the security officer or a 24-hour communications centre to the fact that further action is required.

Depending on the nature of the project, anti-climb alarms on scaffolding could be required, as could alarms around the perimeter, to alert officers or the control room that there is possibly an intruder. Sophisticated alarms can also bring problems. They need mains power – 240 volts – and a secure telephone line. The feasibility of that depends on where the site is and what the circumstances are. If power is generated on site, the logistician will need to ensure that it works at all times. If it works only sometimes, it's worse than useless. Negative or false alarms can also be a problem, especially if directly linked to a response team. Too many false alarms can ultimately impair the speed and level of any response, leaving the site as vulnerable as the boy who cried wolf. There are wire-free and hard-wired alarm systems available, both of which work as remote area alarms. They will guard an open space using wireless radio waves and/or infrared activation.

Finally, it's important to think about whether there is a need to provide personal attack alarms for loan workers. There are other types of loan worker alarms that will work if they fall or collapse because they are gravity sensitive. Given the wide range of alarms available, it is advisable to seek specialist advice to determine what might be the best solution for a particular project.

Special measures

We live in a world where terrorism is an ever-present threat. The average construction site is unlikely to be a target for extremists, but there are a number of high-profile or sensitive construction sites that have a need for special measures of protection. Even if your project is an unlikely target, you should know who your neighbours are. If you are near a potential terrorist target, there might be an attempt to use the site as a launch pad for an attack. High-profile events, such as the Lord Mayor's Show, could also be attacked from a construction site, which could be seen as an ideal platform for terrorist activity. Equally, the target could be the project itself. Animal rights activists, for example, have a history of targeting construction projects and individuals in construction when testing laboratories are being constructed. Also, there are a number of major facilities that constitute a permanent risk, such as airports, docks, military establishments, the weapons industry and the preparations for major sporting events. In the most serious cases, special measures need to be taken from government level down to individual projects because of the magnitude of the risks involved.

In any of these cases, it can become necessary to check for command wires, to ensure that there are specific areas behind hoardings which are clear and that storage areas are marked and patrolled. This technique is called 'target hardening'. It's all about making it difficult for terrorists to use you or get to you. Specialist advice on target hardening should be sought from a specialist security service or consultancy, or from the police.

The concept behind target hardening is rather like recreating the layers of an onion skin. First, the perimeter must be made difficult to breach, by using fencing, technology and active patrols. Then internal security measures need to be put in place. On high-risk sites, patrols will be working on actively preventing surveillance by watching for loitering individuals and vehicles, especially those with any kind of recording equipment.

There is more to site security than ensuring that the perimeter is not breached. Random ID card checks of those already on the site and site awareness programmes to teach staff about suspicious behaviour help to improve on-site security. The risks and issues will be well-publicised on site. Much stricter controls will be enforced that only allow entry to those vehicles that are authorised and signed in. There will be much more attention to the paper trail, and any odd or unusual documentation will be routinely questioned and checked. On very high-risk projects, the threat posed by terrorists posing as delivery drivers can be another incentive to use the construction consolidation centre system, which is discussed in Chapter 11, so that all deliveries aside from the consolidated drop of materials by designated vehicles are removed from the environs of the target.

Despite all this, the reality is that a terrorist with the right skills could get past even the best construction security. The truth is that most components of a bomb are unremarkable and easy to bring onto site individually. This has to be recognised as a fact of the world we live in. What security can do, however, is vastly increase the chances of a perpetrator being caught. All the security precautions mentioned above make a successful attack far harder for the terrorists. In extremely sensitive or high-risk situations, do not be surprised if you find you are contacted by the Security Services, who will want to liaise regarding security arrangements.

This, of course, will not be an issue on the majority of sites. Nevertheless, good liaison with the local police service is an asset to any security operation, and it's worth making an effort to foster good relationships.

References

Considerate Constructors Scheme (2010) Code of practice, http://www.shef.ac.uk/content/1/c6/09/56/61/Considerate%20Constructors%20Scheme%20-%20Code%20of%20Practice.pdf, [accessed 14th January 2010].

Chapter 9
Coordinating Infrastructure and Services

This chapter covers three important and interrelated areas of logistics: catering, temporary accommodation and temporary services. Whilst the permanent building is being constructed, the people on the site require a range of temporary amenities in order for the project to progress. Arranging them well will create a high level of staff satisfaction. Not addressing them, on the other hand, can cause a range of problems and adversely affect production.

Catering

Site catering has evolved greatly over the years. The basic requirements, as defined by the Health and Safety at Work regulations, still only stipulate that staff should be provided with a clean, warm, dry space, facilities to heat a meal and hot water. This constitutes the legal minimum that an employer will have to provide, and there are still sites where that's what you'll get. Standards have steadily risen through much of the industry, however, and on some of the better sites the level of catering can be as good as a West End restaurant. There are sites today where you're more likely to get grilled sea bass and noodles than egg and chips. Catering in construction is becoming more sophisticated, as demonstrated in Figures 9.1 and 9.2, which show the canteen and head chef from the award-winning Unilever House Redevelopment project.

This shift has been influenced by several factors. The growth in corporate social responsibility (CSR) has encouraged more holistic provision for staff, including good-quality meals, but the change has also been driven by the need to be an attractive employer during times of widespread skills shortages. Good food boosts both employee satisfaction and productivity. The recognition that greasy breakfasts do not make for a fit and healthy workforce is also a factor that has encouraged a

Figure 9.1 The award-winning canteen at the Unilever House Redevelopment project

Figure 9.2 Unilever House Redevelopment project's head chef, Kieran Robinson, prepares a lunch meal

change in provisioning. Today's breakfast menu on a leading site will probably include croissants and fruit, and perhaps scrambled eggs and smoked salmon.

Of course, to ensure that the workforce is really happy with its meals, it's sensible to survey the customers and find out what's popular. Apart from anything else, ensuring that the food on offer reflects the range of cultures represented on site can be an important strand of diversity. It should also be remembered that the site canteen is not just a place to eat. It needs to be space which is conducive to relaxation, as it is important for workers to get a proper break during the day. A television might be available in one area for those who want to follow the news or the latest football results, but quieter spaces for reading or talking will be preferred by others.

Canteens, or restaurants, as we now tend to call them, should always include healthy options on the menu. It's a good idea to have educational posters to explain the important of getting 'five a day' (the recommended number of fruit and vegetable portions that need to be consumed for optimal health), backed up by a good selection of fruit and other healthy options. Also, now that many people have food-based allergies, the caterer will have to ensure that there is adequate labelling on foods such as nuts, since these can cause serious allergic reactions.

The modern construction worker is much more likely to be aware of the importance of a good diet and so will not necessarily welcome the choice of the cholesterol-rich traditional all-day breakfast. Their work can often still be dusty and grimy, but they now expect decent provision for changing, relaxation and good food in compensation for this. Catering is one of those operations that trade unions tend to keep a close eye on, to ensure that their members are being properly treated in the workplace.

This improvement in expectations – and thus in facilities – has obvious benefits for site workers, but it benefits the project as well. Apart from the incentive that comes from the opportunity to be an attractive employer, and to have a healthier workforce, the provision of good on-site catering is likely to have a very positive effect on productivity.

If workers have to leave site for lunch, they will tend to take longer breaks – not least because of the extra walking time involved. On top of this, a large number of workers pouring into the site locale at lunch-time may cause disruption to traffic flow and create congestion on the pavements. In a free society, you can't stop workers from lunching where they wish, but providing a quality restaurant makes them more likely to stay on site.

Not only will this reduce congestion, it also lessens the likelihood of staff living up to the stereotypical image of builders through their traditional expressions of appreciation towards young women. Such behaviour needs to be prevented wherever possible – not only is it a poor image to portray, it can be experienced as intimidating or insulting by the targets of this unwanted attention. The only plus side of workers eating externally is the potential benefits to the local economy. Boosts to local businesses are important, but in this instance, the author believes, the potential losses to both productivity and reputation outweigh the gains.

Finally, it should be remembered that there will be different needs at the start and end of a project, when there are fewer staff. In the beginning there will probably be nowhere to accommodate catering. This should be resolved in consultation with the workers. There are now mobile catering units in trucks, which provide hot food to take away. Another option might be a burger van and patio furniture. Film set caterers, who are used to the demands of location filming, could be considered.

Purchasing models

Site catering will often require some kind of subsidy, so that the workforce can be charged less for their meals than commercial requirements would otherwise dictate. Like most contracts in construction, there will be a full scope for tender, which should stipulate what needs to be provided. Numbers to be catered for and the times the restaurant needs to be open are key requirements here.

Even if there are only a few people on site at night and at weekends, it's likely to be prudent to provide them with a similar service to their colleagues who work more sociable hours. Not to do so could constitute discrimination. This will increase costs, however, as even a small sitting will still need a cook, assistants and full electric power. Because a smaller number is being provided with a reasonable choice of meals, there will possibly be more wastage.

This can make for a substantial subsidy being required, although it is not unusual, now that there are more sophisticated and agile catering businesses bidding for these contracts, for the subsidy to be reduced by the practice of selling soft drinks and snacks to increase margins.

Indeed, on a large site, particularly if it is situated in a remote area where there are few alternative places for the workforce to eat, there may be no need to provide any subsidy at all. A sizable, busy site may be attractive enough to induce catering firms to pay for the opportunity to provide the food. This can represent a win/win situation for everybody, since the tender will still make stipulations in terms of the quality and choice of food, and acceptable charging levels.

There are variations on this model. Imaginative caterers will sometimes now offer cash back if they get more customers than expected. Conversely, if the project gets delayed and fewer people are on site than anticipated, the rent may be dropped. Working out the right model is not without challenges, but it is feasible. The caterer can add value to the business opportunity, and to the project, by having a wider range of catering facilities. This could be used to provide sandwiches for meetings and working lunches, etc. Clever caterers will pick up this business alongside their core catering function.

There are several other more mundane issues to consider when creating the scope. Who's going to clean the dining area, for instance? This sounds trivial, but it may be worth hiring a specialist. The area can be partially monitored during the course of the day, but whilst the kitchen staff will clean up any spills of coffee and baked beans, they can't do a deep clean during peak hours. All the time, people will be in and out, many bringing dust and grime from the site with them. This means that the floor surfaces, tables and chairs, restaurant and kitchen area will need regular cleaning. Kitchen staff will probably only be prepared to clean the food preparation areas, so who will attend to the restaurant area and mop the floors? This frequently becomes a niggle on site, and can lead to disputes. Therefore, the solution should be identified and agreed at the tendering stage.

The question of who provides the crockery, thus taking a risk on broken plates, and cutlery will need to be addressed. It should be remembered that on more security-sensitive projects, metal knives and forks may not be deemed acceptable. Plastic cutlery will then be required.

Accommodating catering

Getting the right temporary accommodation for the catering function requires some thought. First, the area will need to be secure. There will be a lot of cash transactions taking place, since site caterers tend not to accept credit or debit cards. This means that there could be a lot of cash on the premises. Ultimate responsibility for this lies with the caterer, but the logistics manager can't just absolve themselves from responsibility for the problem.

Even if the cash is secure, people will often steal the funniest things. Catering packs of sausages and bacon mysteriously disappear. So do various items of crockery. The vast majority of workers will be honest, but sites reflect the entire spectrum of society – like every other walk of life, construction has its criminal element. Therefore, good shutters and doors are going to be an important part of the scope for the accommodation, and the lock needs to be robust enough to withstand a shoulder charge.

When procuring the accommodation, the logistics manager must to think carefully about its functionality. Kitchens have to have extractors, and the air that they extract needs to go somewhere – ideally not the office next door! This is regularly overlooked on sites, particularly when space is at a premium. Fans and fire suppressants will be needed – will they come from the caterers or from the accommodation supplier? This should be checked and agreed early.

There will not always be space on site to create a restaurant with the capacity to house all staff in one sitting. Faced with that situation, the logistics manager may need to devise a system of sittings and make an agreement with the trade contractors about who eats when. However, whilst you can suggest such arrangements, you can't make people eat when they don't want to. It should also be borne in mind that some people need to eat at a certain time for a medical reason such as diabetes. Besides this, it must be remembered that to make appropriate provision for everyone within a diverse workforce there may need to be menu choices that are compliant with the requirements of halal or kosher food.

Organising different sittings makes sense, but is not enforceable. Therefore, it is not always as successful as the logistician would like it to be. Another option on larger sites might be to set up satellite services with scaled-down offerings. If those not concerned about a full breakfast can have a bacon roll by their area of work, that will both ease any overcrowding issues and be another step towards greater productivity – it might easily take 20 minutes to walk to the canteen and back on a spread-out site. However, it must be remembered that satellite canteens need exactly the same hygiene provisions as the full canteen. There must be toilets and a place for workers to wash their hands before they eat.

Canteen culture

Despite all the changes that have taken place in site catering, there is still an ingrained culture around eating in the industry. Breakfast remains the main meal of the day. Typically, workers arriving at seven in the morning will want tea and toast as they're getting prepared to start the day, but a full breakfast will often also be eaten at 9.30 or 10 a.m. In contrast, lunch will often be just a sandwich.

As more and more sites ban eating at the workplace, various initiatives have been attempted to alter the habits of construction workers, but they tend not to be hugely successful. Even if you offer a free breakfast at 8 a.m., there will be little take-up if people are used to eating two hours later. The author has found that people, not unreasonably, generally prefer to wait, even if they then need to pay, in order to eat at a time they're used to. In the author's experience, offering quick alternatives such as fast-food breakfasts also garner little success.

Still, the spurned free breakfasts were a resounding success in comparison with another failed catering initiative, viewed by the author as a learning experience. In order to save space on a cramped site, large restaurant tables that could seat ten, instead of the normal four, were used. A designer was commissioned to create the space and it looked great. However, there was a slight problem: everyone hated it. Construction has a somewhat tribal element, and the trades often tend to socialise together, rather than across the boundaries of their specialisms. In any case, many people do not like sitting down to eat with strangers. It might be argued that tables of ten in the site restaurant could be the holy grail of integrated working, but until that happy day dawns the author recommends creating spaces where everyone can relax with their peers. Cross-cultural socialising cannot be imposed via restaurant design.

In short, the environment has to reflect what most people are comfortable with. It should not be too trendy, and the general atmosphere should convey respect for the workers, not least through cleanliness and quality food. When people are gathered together to eat, it can be tempting to view it as an opportunity to make announcements. Don't. People want to enjoy their food without interruptions. Never forget that this is their time away from work.

Getting the canteen right can make or break the site. If it's efficient and popular, it will do wonders for both productivity and morale. In fact, the logistician will know that it's really good if workers from neighbouring sites will try to get into it. If it's bad, people will vote with their feet, with all the drawbacks to efficiency that this entails. Providing people with good food, like so many tasks within the logistician's remit, can seem a rather mundane function compared to all the innovative and high-tech activities going on. In fact, like many other basic logistics functions, it is one of the core elements that underpin everything else. So it's worth doing it well.

Temporary accommodation

There are a significant number of good-quality specialist suppliers of temporary accommodation. Some of them may only be able to offer a limited range of standard jack-leg cabins, but there are now an increasing number of sophisticated organisations offering a modular build type of cabin which can be adapted to create a bespoke project to reflect the customer's specific needs. There is also a separate class of event-orientated accommodation: marquees, tents, etc. can be procured for marketing purposes and launch events.

Whatever temporary accommodation you are procuring, it is important to understand the purpose it's required to fulfil, and to understand what will be fit for that purpose.

A typical list of temporary accommodation required for a project might include offices, marketing suites, welfare facilities (toilets, showers, drying rooms, changing rooms, locker rooms), security offices (perhaps with turnstiles, as discussed in Chapter 8), reception areas and sentry boxes for traffic marshals and security officers to shelter in during inclement weather. Storerooms, or perhaps even larger temporary warehouse facilities, may be required. And tool hire shops are now a common feature on larger sites, and may need to be planned for.

The quality of the temporary accommodation will also give out a lot of messages, of good or ill, on how much the construction organisation aspires to be seen as a good and caring employer. By the late 1990s, there were only a few sites offering top-class temporary accommodation for the operatives on site, but there are a lot more now.

Respect for People (Constructing Excellence 2010) and the vogue for CSR have not yet achieved global buy-in across the industry, but they have certainly made a difference, particularly with regards to the aspirations of blue-chip clients. This has forced many major construction companies to review the way they think about staff. It's also raised the expectations of the operatives themselves. Professional construction workers now expect good food, decent toilet facilities, a locker and somewhere to hang things up (Figure 9.3). They don't want to wear dirty site clothes on public transport. Nowadays, such changing areas

Figure 9.3 The Unilever House Redevelopment project provided award-winning changing facilities for staff

can be more like the lockers in a quality gym. The time and quality invested in facilities for workers pay dividends. Even in a recession, workers still have some choice over where they work, and in boom times they have an array of options. If facilities are bad, they will go and work somewhere else. Therefore, it pays to demonstrate how much their contribution is valued by putting effort into the elements of the site, such as locker rooms and canteens, that they really care about.

Siting temporary accommodation

The first thing to do, as with so many other aspects of logistics, is to focus on detailed planning. The labour histogram, once again, will be crucial to gauging requirements. It's not just the site staff, of course, who must be considered. It's also important to know what space will be needed for the office-based staff. Their requirements can be extensive. The architectural staff, engineers, construction manager and project manager will all need office accommodation, and the client representative will need at least a hot-desk facility when visiting. The managers of the trade contractors will need space whilst their team are on site as well.

Then, the logistics manager needs to consider the constraints of the site, and work out what space is physically available. Very often, space limitations will be a real challenge. On constrained city-centre sites, you may be lucky enough to be able to acquire a short lease on offices very near the site, but this tends to be the exception rather than the rule. Therefore, the conundrum of how to fit in all the offices for the construction team on site will have to be solved.

Sometimes, imaginative answers have to be found. Sometimes, there is space within the footprint of the building, if there are areas which will ultimately be landscaped, perhaps part of the gardens or a park area. City-centre offices tend to be built to within an inch of the footprint, but other jobs do have features that can be used for siting temporary structures for most of the project.

Very often, however, space is at a premium, and what there is will be needed for both construction activities and for deliveries and access routes. An increasingly popular solution is to use a cantilever scaffold to fit the offices on what will become the first floor. Oil rig-type platforms have also been used to overcome the problem.

Alternatively, accommodation can sometimes be sited in what will eventually become the interior of the structure, but, of course, then you will have to move it sooner rather than later.

All this takes some considerable thinking, and getting it right involves many people from the construction team. The logistician will need to be fully appraised of the phasing of the project, and its implications. It may be necessary to move the accommodation during the course of the build, but the logistics manager will certainly want to mitigate the number of moves. Ideally, you really only want to do it once, if at all, but in practice it may have to move two or three times during the life of the project.

When siting temporary accommodation, it's also important to think about how the buildings will be maintained. From a Construction (Design and Management) Regulations 2007 (CDM Regulations) point of view, as well as for practical reasons, think through whether activities such as window-cleaning can be performed safely.

Sometimes, it just won't be possible to use temporary accommodation at all, and a proper dry-wall office will have to be built, to be stripped out again at the end of the project. Temporary offices can then be put in basements or underground car parks. This will give a good amount of space, but no natural light. Sick building syndrome can happen in temporary offices as much as it can in ordinary buildings, so do everything possible to mitigate the lack of light.

Finally, it's vital for the logistician to make sure that the cabin supplier has got the right measurements. It's a bit embarrassing if the accommodation can't be fitted onto the site. Getting it out can also be a problem. Considering the investment involved, it will be painful to have to demolish the accommodation because, as the project takes shape, it can no longer be got out of its location any other way.

Scoping requirements

The extent of the accommodation can be significant on a major construction project. The office space alone will have to accommodate many different uses and will need spaces ranging from closed offices to large meeting rooms. The largest meeting rooms might need to be able to seat 30 or 40 people, perhaps more, for training, inductions and major meetings. There must also be space for toilets, rest rooms, a kitchen and break-out areas. People need to be able to take time away from their desk, just as they would in a more permanent office suite.

The logistician will need to consider whether an off-the-shelf package of accommodation or a bespoke design would be suitable. Other factors may preclude the purchase of a turnkey solution, besides spatial requirements. The offices may require computer floors or other specialist equipment. Remember when specifying the accommodation that, even for temporary structures, building and planning regulations will still need to be adhered to. When writing the scope, it's often necessary to get help from specialists or suitably qualified colleagues. It would be tragic to buy a package and then find out it needed planning permission and a fire certificate, but it wouldn't be unique in the annals of construction!

If you select the correct cabin supplier, they should be able to advise on all of this. Still, it is helpful for the logistician to be aware of the issues. In addition to the building regulations, the Disability Discrimination

Act should not be overlooked. It's important to stay up to date on all relevant legislation. Any specification which fails to meet current requirements, apart from being bad practice, could incur onerous penalties. Therefore, think these things through at the outset. For instance, if the project needs a cantilever scaffold, it probably needs a lift in order to give reasonable access as well.

The temporary accommodation may have to be erected, in the first instance, in an environment with no electric, water, drainage or other power source. Therefore, the logistician has to work out how the structures are going to be serviced. It's also important to think about waste water and its disposal. Meeting the information communication technology (ICT) requirements can be another difficult challenge. Part of the planning process must involve an assessment of whether it will be possible to install broadband or satellite connections. There is not blanket coverage of the UK at the time of writing, but it's common enough to make it easy to forget that disconnected areas still exist. Will the local system be able to cope with the required number of telephone lines? The ability of wireless signals to travel also varies between areas. Mobile phone signals don't travel well through concrete, and steel will interfere with reception. This needs to be recognised and planned for.

Procurement strategies

Temporary accommodation can be hired or bought, perhaps with a view to getting a buy-back or selling the equipment on, perhaps as an investment to be taken from project to project. Many issues will need to be factored into the final decision.

There was a time when the principal contractor would often just allocate a piece of land where all the contractors could put their accommodation, either within the footprint or on an adjoining piece of land. This did not, however, always lead to the most aesthetically appealing collection of cabins, which could represent a wide range of accommodation in terms of its age and quality. The author's preferred method, which is now more commonly in use, is for the client or principal contractor to procure it for all. There have been attempts in the past to then hire this out to the trade contractors, but this seems like an unnecessary administrative burden when it is remembered that the facilities could simply be reflected in the trade contractor's overall charges.

Accommodation requirements will, to an extent, reflect the contractual and working relationships of the various contractors. Does the team require large, open-plan offices for integrated working, or is it, as in most cases, that there will be higher-status accommodation for the construction manager and client when the trades visit by invitation?

Whatever the exact arrangement, the structure will be a complex building. Arranging delivery will, therefore, be equally complex. Cranes, hoists and road closures will probably be required. If the structure is to be sited at ground level, it will be relatively straightforward to get into place, but even then there will still be work on site to bolt everything together.

This stage has to be approached carefully. The cabin supplier will want to have delivery signed off before anyone starts drilling holes to put in the electricity, water and drainage. Indeed, whether you are renting or buying, there may be limitations to what you can do. Rented equipment cannot be adapted too much, whilst bought equipment will come with a warranty, which, in an ideal world, you do not want to invalidate by putting in extra lighting or shelving. Warranties for the product and structure will likely be separate, and it's important to consider whether the warranty is long enough to cover the life of the project.

Buying right the first time is very important, which is why detailed scoping and seeking advice from specialist suppliers is crucial. On top of this, if you are buying rather than renting, you need to think about what you will do with it later. At the time of writing, the industry is in recession. Many temporary buildings which could once have been easily sold on will now be costing their owners a significant sum of money to store.

Fitting out the offices

The first rule is, despite all the planning involved, to expect the unexpected. The logistician's plans should involve extra capacity and flexible accommodation for unexpected needs. Obviously, if more accommodation is needed because of a change by the client, they can be expected to pay. But it's even better if options such as reusable partitions can be used to create a couple of extra offices in times of need. It may even be that meeting rooms could be partitioned, and just turned back into large spaces occasionally as required. Sometimes marketing suites can be adapted. This is hard on a housing project, as potential buyers will just pop in, but on commercial builds where buyers visit by arrangement there is more scope. The more uses there are for an area, the more the options increase.

Sourcing office furniture may have to be undertaken separately, depending on the specification of the offices. Large companies will have agreements with furniture suppliers, but the logistician will still need to ensure that the items supplied are appropriate. Everything from filing cabinets to fire extinguishers will be needed, just as in a permanent office, and the trick is to anticipate as many needs as possible.

This is particularly important in high level accommodation. It may just be possible to get a photocopier up two flights of stairs, for instance, but who's going to take all the paper up to feed it? These issues sound trivial, but can cause problems in real life.

Office routines will have to be put in place. Who does the washing up and cleans the kitchen area, for instance? Is it someone from the logistics team or will it be done by those who are using it? Will cleaners just be provided for the construction manager's office or will they look after the trades, too? And rules such as taking site boots off before entering the offices will need to be instituted sooner rather than later.

All electrical equipment will need portable appliance testing (PAT), in line with the Institution of Engineering Technology's (2001) *Code of Practice for In-service Inspecting and Testing of Electrical Equipment.*

Of course, whilst this will be an issue in offices, the largest number of electrical devices will be located in the temporary canteen kitchen area. When locating these, particular care must be taken to ensure that there is adequate power, and that there are more rigorous arrangements to deal with fire. It's important to know how much will be done by the caterer themselves, and this should be agreed precisely, but there are specialist kitchen fire suppressant systems, which may need to be brought with the accommodation. Be aware of neighbours also, and whether any fire risk could affect them. When procuring kitchens, remember that there may need to be satellite canteens and toilets on larger sites.

Finally, drying rooms have, traditionally, been provided for workers to change their clothes and hang up any wet garments to dry in heated areas. The last decade or so has seen the provision of lockers for workers, to keep personal protective equipment (PPE) and their personal effects safe. Whilst this is an improvement on the drying room, both end up dirty, litter-strewn and smelly. And the security of personal effects can be an issue, as lockers can be broken into or vandalised. The provision of quality, well-managed facilities is, thankfully, now becoming more common. Case study 7, in Chapter 12, shows what can be achieved.

Temporary services

Temporary water and plumbing

Temporary plumbing services are often overlooked or undervalued on construction sites. Permanent connections are seen as the priority. Nevertheless, to actually make the construction site run, temporary water – in other words the water that is used during the construction phase – is vital for many functions, such as canteens, the cooling of plant and for managing waste water and floods. All water-related needs except those for the permanent structure need to be dealt with by temporary water services.

Like a lot of things within the logistician's remit, the extent of the works required can be quite difficult to estimate. At the planning stage, you will know how big the project will be and what's likely to be involved, but you may not know exactly what is needed in terms of temporary plumbing. As with other functions, such as signage, precise requirements will change over the life of the project.

Reflecting this, a typical scope of works for this service will typically ask for water 'as required'. The package will include water supply to site, removal of waste water from site and responsibility for managing any rainwater or other flooding.

Providing water is not particularly difficult or expensive, but it will need to be configured so that it's available right across the site. As with many services in construction, lump sum contracts are generally favoured. Therefore, an aptitude for accurate guesswork on the part of the contractor is vital. The important thing for the supplier is to try to work out how often and to what extent water supplies will need to be relocated over the life of the project.

Temporary plumbing can either be subcontracted from the logistics supplier or managed directly by the main contractor. In the author's view, it is better off in a stand-alone package than with a logistics contractor, since they will probably subcontract this specialism in any case. Therefore, many of the remarks in this section will be of most interest to a logistician working for the main contractor on the project.

The permanent water supply will be in the core of the building, which contains all the services. The only exception to this is in the sprinkler systems and the pipes in the basement. The core will contain rising mains for use in case of fire. This infrastructure will, on some projects, be a source that the temporary water supplies can connect to. This will not always be possible, but the temporary water supply will still have to be connected to the mains somehow. Similarly, there will need to be a connection to take the temporary waste to the external sewage system. Apart from those two connections outside of the site boundary, everything else will be temporary. All temporary water will be sent through blue piping, so that everyone can see what it is at a glance.

When buying a temporary connection from the water company, you will find that they will want to meter water usage. The temporary plumbing contractor must consider this when tendering, because water rates are now a thing of the past. There will be an accurate assessment of the amount of water used. This is one of the reasons why fitting a stopcock will be one of the first priorities, just in case of flooding.

The main factors in making that connection to the mains water supply will be making sure that there is suitable water pressure upwards in a multi-storey building and creating a temporary ring main. Then, you will need to estimate where you will need water on site and what for. Offices, canteens and toilets will clearly need water, but so will things like wheel washers and jet-washers. There will also need to be taps around site, particularly as a lot of machines need to be water-cooled. Constant running water will be required so that cutting tools can be kept cool. Concrete won't often be made on site, since major pours will be supplied via ready-mix or a batching plant, but those in charge of minor works will need a water supply to make concrete to level an area, for instance, from time to time. Even ready-mix sometimes requires water to deal with spillages. Lastly, some taps or standpipes will need chlorination for drinking water.

It's important to make sure that water pipes are placed where they won't get damaged – they're often best suspended off soffits from beams along the ceilings, although they can be put anywhere that they're unlikely to be run over.

With both water and waste pipes, it must be remembered that it will not always be possible to rely on gravity to keep the pipes moving. Pumps may be required. All this needs to be thought through and installed, and possibly adapted. Maintenance will become necessary, and pipes should be regularly checked for vandalism or damage. It all takes a certain amount of engineering. Besides this, the provider has to work out how to deal with modifications during the life of the site. It's also vital not to forget pumps. There are many anecdotes of flooding in construction where the most memorable part of the story involves someone forgetting to put a drain in: the ability to deal with floodwater is crucial!

Temporary plumbing can connect to an existing sewer, but it may require sump pumps; sometimes you may go down and collect water at the lowest point, pumping it upwards. When constructing basements and lift shafts, they commonly end up full of water. Until there is a waterproof structure, this water is very difficult to get rid of. The construction manager will try to prevent the ingress of rainwater, but, with the best will in the world, it can still get in. The temporary plumbing contractor will need to provide plant and puddle-suckers. These need to be efficient enough to pump for several days at a time if necessary, and need to be available at very short notice. It may be that the contractor can fit a number of channels and downpipes to send water into gulleys rather than downpipes. If water can be channelled, this is usually money well spent to protect against and mitigate flooding. If there is a flood, far more money will have to be spent at a later stage to deal with the problem. Then, as discussed in Chapter 3, it is usually extremely difficult to ascertain who is actually responsible for letting the water in. Given this, it makes sense to avoid the problem altogether where possible.

Non-specialists who are tasked with dealing with logistics can sometimes be tempted to take short cuts to save money, but, paradoxically, these short cuts can prove extremely costly. One project the author is aware of neglected to create any drainage, despite the fact that the site was in one of the wettest areas of the UK, and therefore became waterlogged. Putting temporary pumps in then cost something in the region of a quarter of a million pounds – an expensive, although not uncommon, approach to cost-cutting!

Electrics

In the author's view, temporary electrics comprise such a specialised package that they require a temporary electrics specialist. Skilled as electricians and designers undoubtedly are when dealing with the permanent requirements of a building, there are certain specialist skills in providing the temporary equivalent to a site.

For instance, with designs for permanent electrics all wires have a definite route through the cores and cable trays above the ceiling. Temporary electrics, on the other hand, have to be flexible and agile in order to cope with different ranges of power distribution. The temporary electrical contractor needs to know exactly what power levels are required to make each bit of kit work. Temporary electrics have to be flexible enough to be picked up and moved if needs be.

The power will ultimately connect to the mains outside in the street. Mains distribution units (MDUs), which are big red boxes with plugs in, are designed specifically to power such temporary works. They will only carry 110 volts – a level which can't kill you if you accidentally cut the cable. Permanent electrics can carry a higher voltage because the ring main makes them safer. They can give you a nasty shock, but they will be unlikely to kill you. If you are an end-user of a supply which is not on a ring main, by contrast, you would get the full level of voltage in the event of an accident, hence the importance of providing reduced power levels.

The project will need these MDUs at floor level so that they can be used for tools. The logistician will need to take a view on how many people will be using the 110 voltage main distribution unit, and what they will be using it for. It must also be remembered that tower cranes and some mechanical systems will need appreciably more power.

The high voltage supply needed for the tower cranes will be sent through a steel wire armoured cable. This type of supply will be weatherproof and isolated, and will be needed by trucks and forklifts and temporary accommodation. The canteen will require a mixture of supply levels. Some items will work on a lower voltage, but 240 volts will be needed for equipment such as food-mixers and kettles. Offices need high voltage levels for items such as computers. There will also be activities on site, such as welding, where the power needs are substantial but brief. The system needs to be flexible enough to cope with these short bursts of power consumption.

Temporary lighting will make up part of the brief: as well as being a crucial part of the site safety strategy, it will be used to facilitate work in areas with insufficient natural light. Lighting for staircases and common walkways is particularly important. One of the tasks of the temporary electrical contractor is to provide the power for these facilities.

Activities such as drilling will require task lighting. You might ask contractors to provide their own task lighting, but then you still would need distribution boards, since you don't want everyone plugging into the mains. On a large project, trade contractors may have their consumption of electricity metered and back-charged.

Lighting for hoardings should be considered. Footpaths for the public will often go very near a site, and will need overhead protection against the possibility of objects falling from above. Wooden tunnels are sometimes created for pedestrians to meet this need. These can be dangerous if they are not lit properly, and can encourage crime. Therefore, CCTV and security lighting for these areas needs to go in the scope, which will in any case be pretty extensive. A site may need substantial floodlighting, perhaps so that those working at a height can see the people below. Areas such as loading bays will also need extra lighting.

Good, safe lighting needs to be as consistent as possible. When moving between daylight, dark spaces and artificial lighting, eyes take a while to adjust every time. This should be minimised to avoid an increased possibility of accidents. And don't forget that fire alarms will often depend on temporary power (see Chapter 7 for more details) and that the power source must be totally secure.

With electricity, of course, suitable cabinets for any electrical equipment will need to be weatherproof, and plugs and sockets need to be locked away for safety reasons. The casing needs to be robust enough to stop any budding electricians from the other trades trying to alter the system in any way. Whatever their personal perception of their competencies in this area, electrics must always be dealt with by a specialist.

At the end of the working day, some lights will need to be switched off. This can be done using time switches. Another option might be to entrust this task to the logistics team, particularly the security officers. They will also be able to monitor the pumps, etc. mentioned in the section above. For environmental reasons, things should be switched off whenever possible.

On a construction site, there are steel frames and various other highly conductive materials. Safety around electricity is incredibly important. For health and safety reasons. It's good to wear a dust mask when sweeping up, but it's far more important not to get a potentially fatal electric shock. Therefore, all statutory permissions must be obtained, and maintenance assiduous.

As well as essential maintenance and adaptations, temporary electrics, just like the temporary water supply, will need to be moved as the project progresses. This will require equally refined guessing skills on the part of the electrical contractor. This contractor will be aware that valuable items such as step-up transformers and step-down transformers are prized on site. They will wander. This will often be a case of relocation and reassignment rather than theft. Extension leads go missing with unfailing regularity, although they will often still be in service somewhere on the site. Thefts of cable, by contrast, are becoming an increasing problem. The value of the components is so great that thefts have occurred even when a cable was still live.

Temporary services, both water and electrical, need to be installed properly, and demobilised and stripped out with equal care and attention. Disposal must be both correct and sustainable. Electrical cable is a hazardous waste.

It's also important to remember that construction, as mentioned elsewhere in this section, operates under Murphy's Law. If it can go wrong, it will. Temporary generators always need to be on standby, in case emergency power is required in the middle of a crane or hoist manoeuvre. In the area of temporary services, as in so many other aspects of logistics, you must expect the unexpected and plan accordingly.

References

Constructing Excellence (2010) Respect for people, www.constructingexcellence.org.uk/zones/peoplezone/respect.jsp, [accessed 25 January 2010].

Institution of Engineering Technology (2001) *Code of Practice for In-service Inspecting and Testing of Electrical Equipment*, 2nd edn. IET, London.

Chapter 10
Waste Management and Good Housekeeping

As has been stated in Chapter 3, construction waste is a significant environmental problem. As sustainability issues rise ever further up the agenda, increasing legislation and the rising landfill tax is providing major incentives for the construction industry to review its practices in this area. Therefore, whilst the environmental dimension can make waste a particularly emotive issue, the significant cost involved is now a leading driver for the industry to increase efficiency. And whilst the cost of disposal is usually significant, the good news is that it's usually possible to make substantial reductions in both environmental impact and cost.

Culture change is a long and difficult road, but the industry's vastly improved its track record on other issues, such as health and safety, and there's no reason why it can't do the same with waste reduction.

Still, at the time of writing, there is much work yet to be done. Piecework, as noted in Chapter 4, incentivises wasteful practices. Tasks such as dry-lining will be done as quickly as possible, so a fresh board will probably be cut into every time, rather than using off-cuts. And where manufacturing creates processes that minimise waste, especially since the philosophy of Lean, with its concomitant concern with eliminating waste from the system, construction continues to create bespoke products by building capacity as a way of managing the inherent complexity of its projects. Nevertheless, a review of current practices quickly reveals both opportunities for positive change and areas where things are already beginning to improve.

Procurement of waste management services

There's always a temptation for a construction manager to try to reduce the cost of waste to drive down the price quoted by the waste removal company or logistics contractor. Obviously, it's important to achieve value for money, but it's also important not to lose sight of the fact that it's often far more effective to work with someone who can help you reduce wastage rather than to just opt for the lowest tender.

In fact, the waste management procurement route can often end up as an inhibitor to performance improvement if it is not approached with some thought. Two main methods for waste removal are normally used on major projects. Some contracts are re-measurable, which is almost a form of pay as you go, because the price can be adjusted to reflect the actual amount of waste generated. This does provide an incentive to reduce waste. However, it's still more common for waste removal to be procured on a lump sum contract. This approach doesn't require the logistics waste contractor to be good at waste management, but at guessing how much waste will be generated.

Accordingly, logistics contractors involved in waste removal have become more sophisticated at guessing how much waste a project will produce, based on the type of construction and the track record of the contractors involved. They will take into account historical data from buildings of a similar type, size and level of complexity. Obviously, the length of the programme and the types of materials being used will also figure here.

Although the waste removal contractors will be able to make some assumptions based on previous projects, the current tendering method makes it hard for them to know exactly what they will be dealing with. At the stage at which the contract is awarded, there will possibly be many blanks within the detailed specifications for the project, such as material types and quantities, and the potential for off-site assembly.

For instance, the author once submitted a tender on the assumption that fibreglass would be used for the fire protection, since that had been indicated as the likeliest option. In the event, 25 ml fireboard was used instead. Where the fibreglass would have produced minimal offcuts, which are easily disposed of, the fireboard was dense, heavy and produced a lot of waste. The fireboard was difficult to compact into a skip, and was thus extremely expensive to remove. This is one example of the difficulties faced by waste contractors: they have to put in a price when materials are not yet confirmed, and contingency pricing is often not an option. Another issue with the lump sum route is that no one, not even the construction manager, really knows how much waste a project will generate. If the trade contractors are very conscientious and efficient, the waste contractor profits, but if they are not, the waste contractor ultimately suffers. This makes the whole tendering process for waste a gamble that, it is argued, has no place in a sophisticated modern industry.

What is waste?

This may sound like a simple enough question, but in fact it is getting harder and harder to answer. There is a great deal more to waste than just putting the rubbish in the bin and lifting it off-site in a skip – there are a number of complications.

Technically speaking, waste does not encompass virgin, unused materials. However, due to over-ordering by trade contractors, much of the matter to be removed from the site may be still in its original wrapping. This would not, strictly speaking, require a waste carrier's licence to be removed, but in practice it is likely to be treated as waste. Exactly which materials are waste can become a thorny question – particularly for a logistics waste contractor who is keen to delineate the parameters of a project. One recent project finished with no fewer than 36 truckloads of unused material, which had a value of £250,000, and this was taken back to the London Construction Consolidation Centre. Approximately one-third was taken back by the relevant trade contractor, another third was recycled and the last third went to landfill. The last two-thirds were paid for by the client, both as virgin materials and as de facto waste to be disposed of. This is an important issue, not least because waste carriers have to be licensed in the UK. A more robust definition would be useful, because unused materials, if not defined as waste, should be taken back by the trade contractors for reuse. In practice, however, this is extremely difficult – if not impossible – to enforce.

Much unused 'waste' is bespoke, and it can be difficult, even in pristine condition, to give it away. Typically, it will either not meet specifications or warranties for other projects or be needed at the other end of the country. The transport costs would outweigh the cost of sending it to landfill. This is a pity, because reuse is the optimal destination for waste materials, but even with goodwill and a willingness to offer it at no cost to charitable causes, this can be difficult to achieve.

Nevertheless, it is often possible, particularly in buoyant economic conditions, to find takers for many types of waste. Paper, plastics, timber, metal and drywall can be sold as long as there is a market for the recycled products, although in economic downturns there may not be as much incentive for companies to take them.

Hazardous waste is another complex area, which is subject to the Hazardous Waste (England and Wales) Regulations 2005, amended 2009. Just as defining waste itself can be difficult, deciding whether any given material constitutes hazardous waste is potentially problematic. The Environment Agency recognises three types of waste: hazardous, non-hazardous and material which needs to be assessed (Environment Agency 2009). Various treatments can affect a material's classification. For instance, to the uninitiated, timber is timber. From a waste disposal perspective, however, timber is sometimes hazardous and sometimes not. Materials such as MDF are hazardous when cut, due to the dust, whilst many other types of timber can become hazardous, depending upon what they've been treated with. It's not uncommon to have to dispose of timber from a project in multiple locations. It must also be borne in mind that there are only very limited facilities for dealing with hazardous waste. At the time of writing, there is no hazardous waste tip in the entire country of Wales, meaning that waste from projects in Cardiff and beyond will cost a substantial sum of money to transport.

The debate about what is and isn't waste will continue for some time, or at least until a body such as DEFRA or the Environment Agency have occasion to debate the issue in a court of law. Essentially, at this time, they treat any material that is discarded (not necessarily used) in the construction process and moved off-site as waste. Therefore, you have to be a licensed waste carrier to move it, and whoever receives it must have a waste licence.

On the other hand, some waste is straightforward to deal with. Plastic wrappings and cardboard can be baled and taken away on pallets. This is not difficult, but must be done regularly so that the material does not constitute a health and safety risk. Building in processes to ensure that this happens efficiently is not always straightforward, but with determination improvements in performance can be achieved. In particular, this is where access to a construction consolidation centre (CC) licensed to handle waste, or a dedicated waste CC, would be a huge advantage. Otherwise, achieving the necessary quantities to justify removal in a timely fashion can sometimes be difficult.

Challenges to good housekeeping on site

As stated earlier in this chapter, tendering for waste removal contracts can be about being good at guessing how many skips will be needed. Another crucial part of the procurement process is creating lines of communication between the contractor and the construction manager so that both sides are clear about what can be expected. Scopes of works tend to be loose in parts, and it is important for the principal contractor to be clear about what they are trying to achieve.

A common problem is for the waste company to be contracted to provide and empty wheelie bins, which the trades are expected to use for their rubbish whilst on site. Therefore, if the trade contractors throw rubbish on the floor, it's not the job of the waste contractor to it pick up. Despite this, if there is a lot of rubbish lying around, this is generally perceived as their dereliction of duty. There tends to be different obligations for different contractors. It's not unusual for the reinforced concrete frame contractor or block- and brickwork contractor to have to dispose of their own waste. However, they have been known to use the wheelie bins provided for others as a way of saving money. This causes conflict and safety issues as the logistics waste contractor should not be paying for their waste. In any case, wheelie bins are not designed for heavy waste.

The manager in overall charge needs to be clear about the answer to two questions: 'Who is responsible for putting rubbish in the bin?' and 'How are we going to manage rubbish that is not in the bin?'

Trivial as it sounds, this is an area that causes huge arguments, and all waste contractors have to deal with this issue on construction projects. Whilst expecting contractors to bin their own rubbish is still common, it tends not to work well. Asking them to put the right rubbish in the right bin is even more difficult.

It creates an extra area of confrontation for the construction manager, who will want to minimise the number of disputes on the project. This impulse can stem from a laudable commitment to integrated working, but it can sometimes be caused by a lack of confidence in dealing with disputes. While few would mourn the passing of the old, macho site culture, we are currently in a period of transition which can cause real problems, especially for younger construction managers. The new breed of university-educated managers lack the years of informal site-based mentoring that was enjoyed by their predecessors. They may have had few opportunities to hone their negotiating skills (Collier 2009a). However, there are still plenty of robust individuals from the old regime on the average site who may not fully appreciate the softer skills deployed in a more conciliatory approach. At worst, experienced trade contractors can effectively bully the package manager, particularly if that manager is only recently arrived from academia. Even at best, the busy manager of any site, however solid their negotiating skills, will not be looking for disputes. The issue of who should have put the wrapping in the bin, therefore, may very well be too low down the list of priorities to be tackled.

Even if the construction manager is prepared to insist that trade contractors pick up after themselves, it is not always easy to work out who has failed to clean up. Some packaging materials, for instance, could have been discarded by any one of several contractors on the average project. Even if the culprit is clear, the traditional method for addressing the problem – the issuing of a corrective action notice – is not generally recommended by the author.

A corrective action notice will usually give the contractors 24 hours to clear away their rubbish, with the alternative being that it will be cleared up for them and they will be charged. This sounds good in theory, but the rubbish then has to be photographed and it will have to be proved that they have been given a full 24 hours to clear it. Still, there is likely to be a dispute, and the resultant charges will often be dropped during the final account negotiations. Ultimately, either the principal contractor or the logistics waste contractor is left with the bill, so nothing has been achieved.

What has proved more efficacious is to instruct trade contractors that they will get no more materials delivered until their waste is cleared, on the grounds that a messy site constitutes a health and safety risk. Then, they will generally clear it. Even so, this is still not the most efficient approach to waste removal that could be chosen. By opting for a more comprehensive solution to the waste issue in the first place, many problems can be avoided.

A professional approach to waste management

As has already been stated, legislation relating to waste management is on the increase. Underlying this are both environmental concerns and increasing standards in health and safety. It seems likely that, in the not too distant future, waste handlers will have to be trained to spot hazardous waste before they pick it up. Using a general labourer may no longer be acceptable in the future.

Even without that, there are environmental drivers which increase the complexity of the task. The hierarchy of waste (Figure 10.1), whereby prevention of waste is most favoured and disposal is the least preferable option, creates a blueprint for better waste management, but it also creates complexity.

The days when all waste could be thrown indiscriminately into the same skip and sent to landfill are going, and are very unlikely to return. Every site must have a site waste management plan (SWMP), if the project is located in England and the build will cost more than £300,000 (Price *et al.* 2009). Not only are there statutory requirements for change, there are also schemes designed to help leading construction stakeholders excel in their performance. At the time of writing, a good example of this would be the Waste and Resources Action Programme (WRAP), which has initiated a campaign to halve waste to landfill. This initiative, inspired by successful programmes in the retail sector, gives companies a chance to make a public declaration about their performance targets. According to Watson (in Collier 2009b), the head of construction at WRAP, whilst a 50% reduction in waste is a tough target for many small- and medium-size enterprises (SMEs), leading contractors are setting themselves ever-more challenging targets, some expecting to

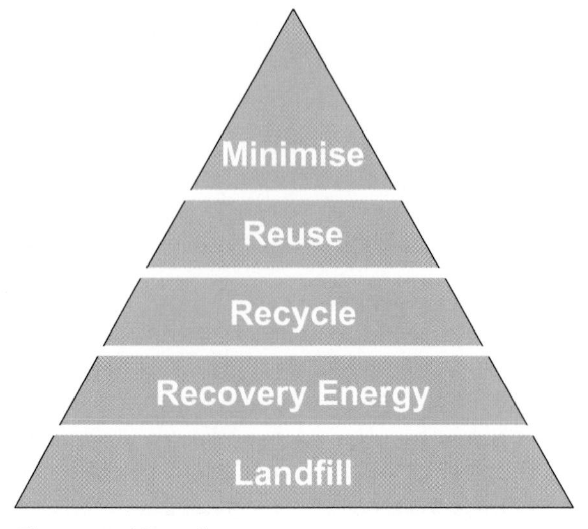

Figure 10.1 The waste hierarchy

achieve zero waste to landfill in the near future. Construction can be competitive, and Watson thinks the peer pressure that breeds competition between construction companies will be a major force for positive change in this arena. There has never been a better time to take waste management seriously.

One of the changes this entails is that greater segregation of waste is required (Figure 10.2). This, as well as likely future legislation, creates a stronger case for hiring specialist waste handlers. This team does not have to be the same as the specialist materials handlers, but it can easily be part of a multi-skilled logistics team. There are several advantages to this.

Most importantly, perhaps, it can create a team ethos. Removing waste is generally not perceived as a high-status role, and creating a team which rotates roles both mitigates the repetitive nature of the task and improves the way the workers at the coalface see themselves.

Figure 10.2 The range of waste segregation bins that can be used on construction projects

The best way to approach motivation for such roles is to emphasise both the importance of the role and its customer service dimension. It's easy to forget how many of the staff in such teams, whether there through choice or circumstance, are bright, articulate and enthusiastic about their work. If you explain the value of their contribution, and defend them from any pejorative remarks from others on site, you can make a huge impact on their commitment and performance levels. Highly skilled trades are all very well, but someone always needs to tidy up and make the tea. Both contributions are a vital part of the process, and if the waste disposal role was afforded more respect and understanding, it would be likely to attract more motivated people.

Specialist contractors are more likely to have invested in specialist equipment to make housekeeping tasks more efficient. There are tools out there which can allow one operative to do the work of perhaps ten people equipped only with standard brooms. There is a tradition on construction sites of relying on muscle power, but the right tools can make the role far quicker and easier to perform.

Another crucial part of the equipment that will need to be thought about is the waste bins. The logistician, if in charge of the waste function, needs to establish the right number of bins to place on site. There's no one-size-fits-all solution to this. It will very much depend on the size and layout of the project. Still, there is one unchanging rule: however many bins you think you need to service the project, you actually need 25% more. Full bins have to be emptied, and they need to be replaced with new bins as they are removed (Figure 10.3). The size of the bins

Figure 10.3 Waste operatives at the Willis Tower fit out project moving waste from the workface

required is another consideration, and this may change through the life of the project, depending on the number of trades on site, and the activities they are performing. It's important to get bins that will fit through doorways, and in and out of lifts and hoists etc. as required. Waste removal needs to be coordinated with site-wide logistics resources so that hoist, crane and delivery bay time is allocated for delivery (i.e. reverse logistics).

Skips are a common waste removal tool, but one that is seldom used efficiently. They are charged for on the basis of their full cubic capacity, which generally means that a large proportion of the charge is for removal of fresh air – there is generally plenty of space between the waste materials, and they are often removed from site when half-full purely because they're taking up too much space. That's why compacting waste materials has been common practice since the 1980s, at least on larger sites, where skips have often been replaced by dust carts. On average, they have a 40 cubic yard capacity, which roughly equates to 100 cubic yards of uncompacted waste. More volume means fewer vehicle journeys, but there will be no reuse and little recycling.

Potential for further improvement

As discussed in Chapter 3, construction companies sometimes effectively fudge the issue by quoting the generic recycling rates of the transfer stations they use. Furthermore, even if a high percentage of waste is diverted away from landfill, if the total waste is high (and in construction it generally is), that still leaves a significant amount of tonnage going to landfill. All waste reduction is good, but until the industry gets to zero waste to landfill, preferably by creating less waste in the beginning, we shouldn't congratulate ourselves too much.

In an ideal world, more far-reaching changes need to be made to the way the industry deals with waste. More use of off-site manufacture could reduce the amount of waste created, both in terms of offcuts and of packaging.

There is also a huge amount of scope to reduce waste creation through the use of construction CCs, which are discussed in Chapter 11. A recent WRAP case study has demonstrated the ability to divert 84% of waste away from landfill by using a CC (Lundesjo 2009). The author has been involved with a project which achieved a 99.24% reduction of waste to landfill. A dedicated, specialised team can then deal with the segregation and reclamation of much waste before it reaches site, as well as reducing wastage through damaged materials and over-ordering. Moreover, a specialised logistics team can do far more than just deliver and empty bins. Through the rationalisation of support services into one team, the capacity to provide bespoke responses to waste creation can be made possible. It also frees up the trades to do what they do best, perhaps even simply texting to request logistical support with waste disposal as needed.

Rationalisation of waste services can create huge savings. One might imagine that the possibility of a six-figure cost reduction might get senior attention. However, perhaps because waste management is seen as so intrinsically unattractive and uninteresting, it rarely gets the attention that the size of the potential savings ought to command. Too often, waste management gets left to demoralised temporary staff equipped with only brooms, shovels and skips. This area therefore seems too marginal and trivial to focus on with the kind of value engineering which is applied to other facets of the construction process.

Housekeeping

The very nature of construction ensures that a substantial amount of mess is created during the life of a project. Many people will be on site, and most of them will be cutting and fixing various materials, some of which will be immediately discarded. Dealing with waste removal has been discussed, but it should be remembered that, alongside the discarded materials, the sheer numbers of people and the intensity of activity taking place will create an entirely separate class of waste material. Discarded soft drink cans and polystyrene cups intermingle with other general litter. There is also likely to be a huge amount of dust. Rainwater may well need dealing with at various points in the project. And all of these issues will have to be dealt with if the site is to be a well-ordered working environment.

People have their individual attitudes to tidiness. Some are naturally more organised than others are. Unfortunately, the focus on completing a specific task militates against any impulse to stop and tidy up for many workers. Also, some craftspeople will see it as simply 'not their job' to tidy up. Once an area is already a mess, most people will then find even less incentive to deal with their individual litter. Even if you try to get people to clean up their own mess, as we have already discussed, working out who is responsible for it is no simple task. It's often very hard to work out who is most likely to have left discarded offcuts of material. Since banana skins and soft drinks cans have no known association with any particular construction trade, finding the culprit is likely to be next to impossible. This is not only aesthetically unpleasant; it can also quickly become a serious health and safety issue.

Welfare areas can prove as much of a problem as the workface. Workers have to be given somewhere to change and dry their wet clothes. Whilst they're there, they may well eat their sandwiches or read the paper. Discarded food wrappers, newspapers and half-eaten fruit create a depressingly unkempt area extremely quickly, and there is then little likelihood of the next occupants taking it upon themselves to do much to improve the situation. Because litter affects everyone who uses a space, the lowest common denominator can so easily prevail. In an ideal world, everyone would clean up after themselves. In the real world, by contrast, it seems sensible to pay someone to keep on top of the situation, rather than leaving rubbish to fester until the almost inevitable frantic attempt to clean up before a VIP visit.

The overarching issue with housekeeping is that once the rot sets in bad behaviour gets worse. Everyone takes a 'no one else cares, so why should I bother' attitude. Therefore, the principal contractor needs to manage this as a separate issue from waste. As multi-service gangs become a more common solution to logistical issues, there is once again a choice between unskilled agency labour, which will need proactive supervision, and the multi-skilled specialist logistics team. This service can provide a solution to the problem of maintaining consistent standards by taking general responsibility for common areas such as walkways and stair cores. Since these are areas that people are moving through regularly, there's an important safety dimension to ensuring that these areas are kept clean and tidy. Both staff and visitors will be less likely to trip and fall.

For the logistics team, their biggest challenge will come from the raised expectations this will create. If the staff become used to higher standards, they will start to complain about minor littering, where once they would have accepted complete chaos. This increases the pressure on the team, but it has to be seen as a positive effect. People who take more pride in their place of work will do a better job.

Sanitary facilities

Toilet cleaning is a particularly important task on site, however unglamorous it may seem. The state of the sanitary facilities will have a huge effect on morale – either positively or negatively. Clean toilets are also important from a health and safety perspective, since poor facilities are a health risk.

Anyone entrusted with the task of cleaning the site toilets needs to be properly trained and provided with full personal protective equipment (PPE). The training needs to encompass the Control of Substances Hazardous to Health Regulations (2002), since the cleaning materials will be hazardous substances. Too often, risk assessments deal with hazardous materials like glues and mastics and overlook the bleach for the toilets! In terms of PPE, the operative should be equipped with strong rubber gloves up to the elbow, wellington boots, goggles and a mask. They need to be able to cope with what can be extremely offensive graffiti and possible vandalism. The author always argues for extra financial compensation for the toilet cleaners due to the unpleasant nature of the working conditions. The industry has worked hard to create easy-to-clean toilets, but they are not popular with staff, since they are essentially stainless steel boxes. The author would instead argue in favour of a permanent attendant in the toilets, to stop facilities being abused, but it would take an individual with a certain resilience and force of personality to enforce such standards. That said, the workers themselves will sometimes become the de facto police for the facilities, informally resolving problems. Lastly, money invested to create first-class facilities will be rewarded with better behaviour.

Another consideration is whether to provide electric air dryers or paper towels. A lot of workers like to have a proper wash down at the end of the day, and cannot dry their torsos effectively with electric dryers. Towels, however, create waste. Therefore, it's a good idea to provide both options. The bins will need to be emptied regularly, since they will fill up very quickly. This may sound trivial, but in fact there is often a huge cost to disposing of the paper towels. Since this can be expensive, it is sensible and worthwhile to explore ways of collecting them so that they can be recycled.

The facilities must be planned and resourced in a way that reduces opportunities for theft. Without a lockable toilet roll dispenser, such as are found in public toilets, toilet rolls will disappear in droves. Similarly, the degreasing hand cleaner should be placed in secure dispensers, which need to be both robust and primed to dispense slowly, so that it becomes inconvenient to stand underneath it with a jar. In fact, all equipment has to be robust and able to withstand frequent – and sometimes heavy-handed – usage.

Achieving good housekeeping

The phrase 'sweeping up' always conjures up the image of a broom, but using one is actually very time-consuming and inefficient. Perhaps 50% of the dust is just swept into the air (and people's lungs) before settling elsewhere. Sweeping dust is something we, as an industry, need to wean ourselves off. In a concrete stair core, coming down perhaps six levels or more, sweeping down from the top can create a huge amount of dust. Mechanical and electrical tools should be used wherever possible. Unfortunately, vacuum cleaners and floor sweepers that are robust and reliable enough to survive on a site are very expensive. Even the average industrial vacuum cleaner will only survive a couple of weeks on a site. Construction dust is heavy, and the cleaner will be in constant use.

However, push-along sweepers are a viable alternative to reduce dust in the air (Figure 10.4). Ride-along sweepers are another option, although the model needs to be small enough to get through doors. Likewise, it needs to be light enough for the capabilities of the floor plate. These small vehicles are often very heavy, and the weight is extremely dense due to their small size. Also, these machines often struggle to cope with

Figure 10.4 A logistics operative uses a push-along sweeper during construction of the extension to the Manchester Arndale shopping centre

the uneven surfaces of an unfinished floor. It's still very difficult to clean rippled concrete. Brooms cannot, therefore, be done away with altogether, although their use can certainly be reduced.

Good housekeeping also involves the management of activities such as drilling or cutting. If workers are using angle-grinders on paving stones and the like, it's important to ensure that the dust is not flying into the faces of the workers or exiting into the street and covering the public.

These issues all sound like minor considerations, but in practice they can take up a fair chunk of the logistics manager's day. Still, these are tasks with no direct effect on productivity, so an effective regime needs to be put into place. When the construction manager is buying the logistics package, housekeeping can be another grudge purchase. There may be some debate about how many people are needed to clean up. Since housekeeping can become so labour-intensive, the author views investment in equipment to be money well spent. Still, there needs to be a realistic view regarding what can be expected from the logistics contractors and their housekeeping team. It can't be a spare-time job – there's too much to do. It needs to be a coordinated and planned function, where individual operatives take responsibility for different zones and levels.

As discussed in Chapter 4, the logistics team will have to perform some of the most menial tasks on a construction site, but their contribution is vital to the entire construction process. New managers in construction, conversant in both the most sophisticated management theories and the incredible potential of engineering, often take time to understand that these everyday functions are just as important to a successful project as the design of the building. The toilet cleaner and the architect are both indispensable!

References

Collier C (2009a) *Building Visionaries*. Chartered Institute of Building, Ascot.

Collier C (2009b) Halving waste to landfill. *Construction Information Quarterly* **11**(1): 31–2.

Environment Agency (2009) What is hazardous waste? http://www.environment-agency.gov.uk/business/topics/waste/32200.aspx, [accessed 20th January 2010].

Lundesjo G (2009) *Construction Logistics Key to Performance at Central St Giles: WAS610 – Material Logistics Planning Case Study*. WRAP, Banbury.

Price T, Wamuziri S and Gupta NK (2009) Should SWMP 2008 be integrated with CDM 2007? *Construction Information Quarterly* **11**(1): 12–17.

Section 3
The Future of Construction Logistics

Chapter 11
Construction Consolidation Centres

This chapter introduces the consolidation centre (CC) concept and methodology and provides a compelling argument for applying it to the construction industry based on its successful use over many years in the retail and manufacturing industries.

Associated and integral supply chain management (SCM) techniques such as warehouse management systems (WMS), barcoding, value stream mapping (VSM) and radio-frequency identification (RFID) are also introduced to give an insight into the more sophisticated goods procurement and distribution techniques that facilitate better logistics management.

The first example of a dedicated logistics facility for a major construction project in the UK is the CC at London's Heathrow airport. This innovative facility was the subject of a Department of Trade and Industry (2004a) research project undertaken by the authors to investigate the potential for transferring the consolidation centre methodology (CCM) to other areas of the construction industry.

The social, economic and environmental benefits of utilising CCs are significant and are aligned with the increasing importance and implementation of corporate social responsibility (CSR), which is also introduced in this chapter.

The CCM is applicable to any type of construction project and there is no one-size-fits-all model. Three CC options are described in this section that can apply to different project circumstances to maximise logistics efficiency:

- concealed consolidation centre
- communal consolidation centre
- collaborative consolidation centre.

Consolidation centre concept

CCs are essentially buffer storage facilities that hold materials or goods for a limited period prior to their being required by a retail outlet or site, unlike warehouses that may store materials for long periods. CCs are normally situated strategically near to motorways or railway stations to facilitate ease of delivery and hold vast quantities of materials or goods in purpose-built storage racking. They incorporate sophisticated computerised WMS to receive, locate and despatch materials and goods using barcode and RFID technology. CCs have been used successfully in the retail sector for many years but it is only recently that they have been adopted for use for construction projects. The earliest example of a CC being used to support construction projects was that established by British Airports Authority (BAA) at London's Heathrow Airport in 2001. Coincidentally and independently an almost identical facility – the Logistik Center – was being developed at the same time in Stockholm, Sweden to service a major urban regeneration scheme. The two projects could not be more different from one another but the CC concept and methodology applied was identical: to maximise logistical efficiency whilst minimising disruption to the resident stakeholders.

Improving the materials delivery process to site

Whilst the modern approach to logistics has improved project performance *on* site, it has had little effect on the flow of materials *to* site. The delivery of materials and equipment to site is a process that often has a considerable adverse effect on other stakeholder groups, yet remains the least controlled and coordinated element of the construction process. At a time when social responsibility is moving higher up the corporate agenda, with many clients aware of the significance of public perception, and when the industry is under sustained pressure from government to improve its image, this is unacceptable. A more coherent strategy is therefore required, one which improves logistics techniques throughout the entire supply chain. To achieve this, some method of improving the supply chain process before materials and equipment arrive on site must be considered. The CCM places more emphasis on all parties involved in the construction process to plan and coordinate their activities more effectively.

The Heathrow Consolidation Centre (HCC) was set up to provide an alternative method of delivering materials to site. Its purpose was to promote the efficient flow of construction materials, plant and equipment from the supplier to the point of use on a project.

Figure 11.1 illustrates the HCC's process of receiving and delivering materials.

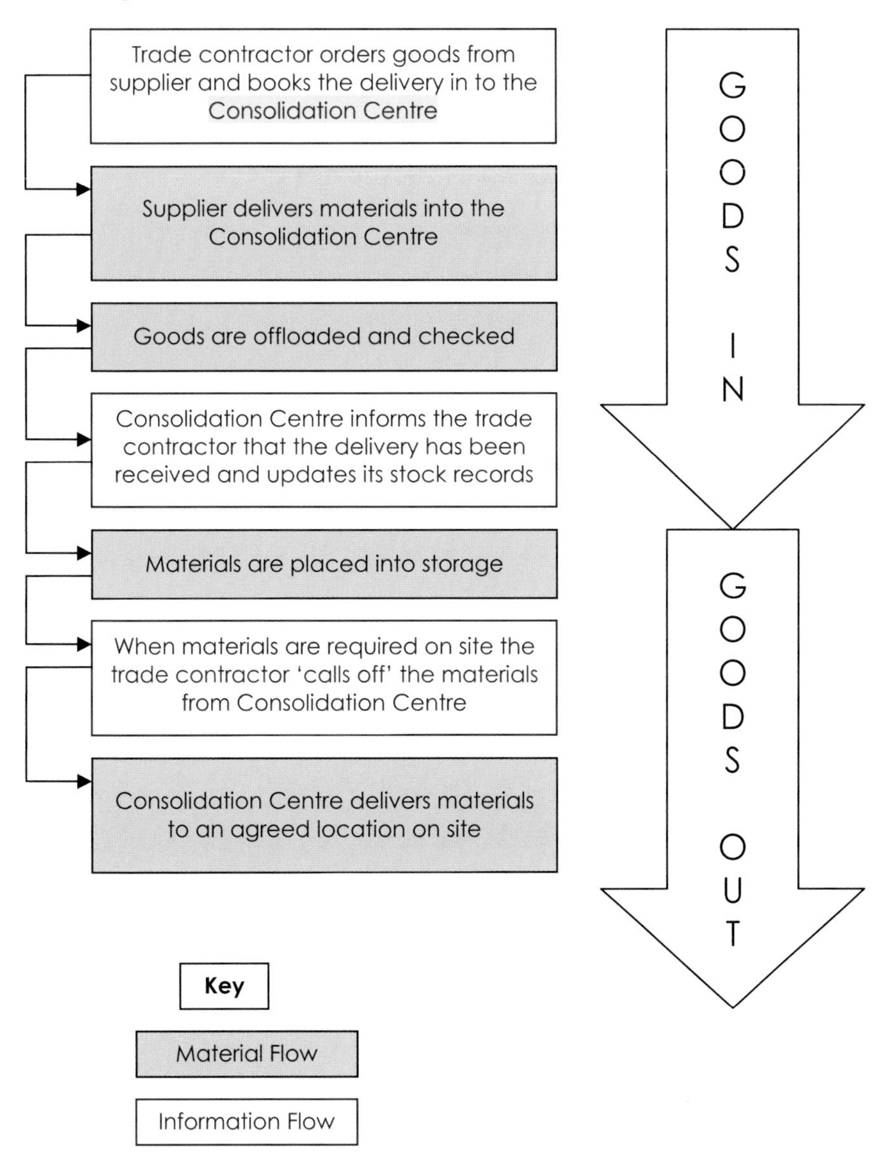

Figure 11.1 The consolidation centre's process of receiving and delivering materials

Based on models which are common in both the retail and manufacturing industries, the HCC overcomes many operational difficulties associated with construction logistics by:

- reducing the numbers of individual deliveries of materials and equipment to site by consolidating consignments

- providing the option of delivering materials at night when roads are less busy
- providing a more controlled and accountable delivery service by using the single point of contact the CC provides
- reducing harmful CO_2 emissions by using appropriate low-emission, fuel-efficient vehicles
- delivering materials to site in the exact quantities required as determined by the predicted shift productivity
- maximising the productivity of the skilled workforce by eliminating the need for them to handle materials
- reducing the potential for manual-handling-related accidents and injuries through the deployment of a properly trained and equipped materials handling workforce
- increasing time reliability by acting as a production buffer for materials with extended lead times
- reducing waste by using reusable delivery cartons and packaging
- influencing improved logistics techniques further up the supply chain.

Construction logistics consolidation centres: an outline

This section provides a basic outline of the CCM based on its application at London's Heathrow Airport (airside projects).

Basic functionality of a consolidation centre

The operation of a CC involves the receipt, temporary storage and distribution of construction materials and equipment.

Materials and equipment are delivered to the CC by the supplier or haulage company, stored for a short period (in a warehouse environment located away from site) and delivered to the point of use (on site), on a just-in-time (JIT) basis by a dedicated logistics team working from the CC.

A CC should operate a palletisation strategy to ensure safe materials handling and storage throughout the delivery process. This means that all consignments arrive on a pallet or in a stillage. A time limit should also be imposed in relation to the storage of materials (this should ideally be set at 10 to 14 days) to ensure the efficient use of storage space.

The application of the consolidation centre methodology

The application of the CCM is very flexible. A CC could be set up to serve a single project, multiple single-contractor projects (in a region) and even multiple contractor projects in a collaborative agreement. The concept can easily be tailored to suit individual site requirements and it may be operated by a principal contractor or a specialist logistics contractor, depending on the size of the CC, the size of the site it serves or the number of sites it serves.

The types of materials handled

A CC is capable of handling the vast majority of materials and equipment used on a construction site (with the exception of time-critical items such as ready-mixed concrete). If the CC is to handle heavy items such as steel-frame components or precast concrete members, it must be resourced appropriately. For this reason, it would be appropriate to impose a weight limit for materials to suit the lifting capacity of forklift trucks etc. to ensure the efficient and cost-effective use of mechanical resources. Any materials that exceed this weight limit would need to be delivered directly to site.

Consolidation centre building

The location of the CC is very important. Ideally, it should be close to the strategic transport network (to ensure convenient access and egress of delivery vehicles) and should be no further than 10 km from site. Other requirements include:

- a warehouse for the storage of perishable materials (this facility should include pallet racking to maximise storage space)
- an external compound for the storage of non-perishable materials
- a covered delivery bay for the receipt of materials during wet weather
- an office facility for administration
- welfare facilities for the CC employees and delivery drivers.

Mechanical resources

Mechanical resources at a CC will typically include:

- a forklift truck
- a number of hand pallet transporters
- a delivery vehicle (e.g. 27 tonne rigid bed lorry); flatbed lorries are ideal and should be fitted with a mounted crane to ensure the efficient offloading of materials on site (even if there is no forklift on site).

Information communication technology

Generally, ICT resources will include:

- basic WMS: a simple paper-based system might be sufficient depending on the size of the CC and the nature of the project
- barcoding or RFID facilities
- computers for maintaining database records and for communicating with suppliers and contractors via email and the Internet.

The benefits of using a consolidation centre

Many logistical constraints associated with construction work can be managed by the operation of a satellite CC (located away from site). Situations that would benefit most from a satellite CC include:

- Inner-city construction projects where the lack of storage space and facilities forces materials to be delivered on a JIT basis. This level of sophistication simply is not practicable using traditional logistics techniques.
- 'Live' environments where construction productivity is made around the operational requirements of the client such as railway and airport operators, schools, hospitals and retail businesses etc. that need to remain open throughout the duration of construction work.
- Security-sensitive sites, e.g. airports, courts of law or military facilities.

Social and environmental benefits

Operating a CC reduces congestion on and around the site by reducing the number of construction-related vehicle movements. An investigation into the Heathrow CC revealed that its use reduced construction vehicle movements by up to 50%.

Using traditional methods, sites often manage their resources (such as delivery bays and forklift trucks etc.) by allocating suppliers with a delivery window. However, the investigation found that this policy had a negative effect on the environment as the majority of suppliers said that they meet their allocated times by arriving up to an hour early and parking outside the site or driving around the roads nearby. This, however, causes more congestion, disruption and vehicle emission pollution. The investigation also found that many vehicles were severely under-utilised as a result of the reactive, client-led demand delivery strategy operated by the suppliers.

In addition, the CCM mitigates many adverse practices associated with traditional construction logistics by:

- Enhancing control over the delivery process. The single point of contact provided by the CC makes it easier for the site to make short-term changes to the delivery schedule in response to occurrences on site or in the vicinity of the site.
- Making it easier to impose off-site traffic management systems, e.g. to ensure that large vehicles avoid residential areas.
- Making it easier to enforce policies such as restricted delivery times, e.g. to avoid clashing with rush-hour traffic.

Project performance benefits

The CCM enhances time reliability as a result of:

- Increased productivity: this is achieved by eliminating the need for skilled tradespeople to leave the workface to unload equipment and materials. Traditional site practices usually involve the skilled trades workforce in the process of receiving, handling and storing of materials, which inevitably results in lost production time.
- Guaranteed production buffer: the period when materials are held at the CC reduces the potential disruption caused by late deliveries and avoids the need to store materials in poor conditions on site for long periods, which might result in their being damaged. This is particularly useful for materials which are manufactured overseas or which have a particularly long lead delivery time.
- Reduced risk of delay and additional costs caused by damage to materials: these often occur as a result of poor storage conditions on site or poor handling techniques employed by site personnel.
- Improved standards of safety on site: as a result of using a properly trained and equipped workforce to handle all the material distribution operations, safety standards are increased. Sites served by a CC are generally tidier, as there are no materials left around to obstruct the workface.

Future developments

In the future, the CCM will seek to:

- enhance vehicle utilisation rates by incorporating a waste management strategy into the operation of a CC
- promote good SCM practices – it is possible to reduce the amount of packaging waste by using reusable delivery containers with the suppliers because regular deliveries will be made and careful offload,

storage and distribution techniques will be employed by the logistics operatives
- increase efficiency by encouraging the design of a work pack
- use barcode technology to improve the flow of materials throughout the supply chain
- use RFID technology to improve logistics and also benefit the building post-construction stage.

The challenges of pricing the logistics element of a contract

It is difficult to accurately quantify the cost savings which occur throughout the supply chain as a result of a CC. It should be possible to extract some of the logistics and associated costs at tender stage; however, proving that materials suppliers, haulers and trade contractors are benefiting financially is difficult due to the level of uncertainty which exists when pricing the logistical element of a contract. It has been demonstrated that suppliers and hauliers benefit from shorter vehicle and driver turn-around times when delivering to a CC compared to delivering to a construction site and that trade contractors benefit through more efficient use of their workforce. The client ultimately benefits financially through efficiency gains, materials delivery certainty and lower waste costs.

Even though the Heathrow CC has demonstrated that it is possible to provide this form of logistics service and deliver a high standard of performance (e.g. getting the right materials to the right place, at the right time, in the right condition and in the right quantity), the client or the operator of the CC must take responsibility for, and insure against, damaged materials and the repercussions on site. It should be possible to extract the risk contingency at tender stage, but this is difficult because of the level of uncertainty that exists when pricing the logistics element of a contract.

Change management

Construction industry practitioner's resistance to change and cultural factors like those identified previously in Figure 2.7 are likely to inhibit the successful implementation of the CCM. A prolonged consultation period was therefore undertaken with the trade contractors, construction managers and suppliers before the HCC began operating.

The consolidation centre methodology

When the original CC was established at Heathrow Airport in November 2001, many claimed that it was a specialist facility and that this approach to logistics would only be applicable to airport construction projects that have stringent security requirements. Just four years later, the London Construction Consolidation Centre, based on the same principles as the Heathrow model, was set up by Wilson James to serve four projects in central London. Both CCs had a different set of circumstances which made it necessary to consider alternative means of delivering materials. At Heathrow, the main drivers for the CC were the:

- high level of security required at an airport
- need to deliver major construction projects without adversely affecting the primary operational needs of the airport, e.g. the efficient processing of passengers
- the commercial needs of the airport to require that project handover dates are essential.

In central London, the main drivers for the CC were associated with:

- limited space to park and offload materials on site
- severely congested access and egress to and from the site
- the financial burden of the London congestion charge.

Additional factors which were discovered after the HCC was founded and acted as a driver for the London Consolidation Centre include:

- increased productivity on site
- cost savings as a result of faster turn-around times and better use of suppliers' delivery vehicles and drivers
- reduced damage and waste as a result of using better logistics techniques
- improved health and safety standards as a result of employing specialised logistics operatives and techniques.

Heathrow Consolidation Centre: Processes, systems and technology

The punctual availability of materials and equipment is essential for any construction project and using and integrating robust systems, processes, equipment and ICT is essential to ensure the CC's ability to deliver a reliable service.

To facilitate effective communication, a CC would use appropriate documentation incorporating a standard set of forms. These forms are used to record and transfer the information necessary to support the processes of receiving, storing and dispatching materials. Figure 11.2 illustrates the materials-in booking form used at the HCC. This form captures full details of the description of materials, suppliers' details, offloading arrangements and time taken to offload to be referred to for future performance-monitoring purposes. Trade contractors should provide the information required in the 'materials in' booking form at

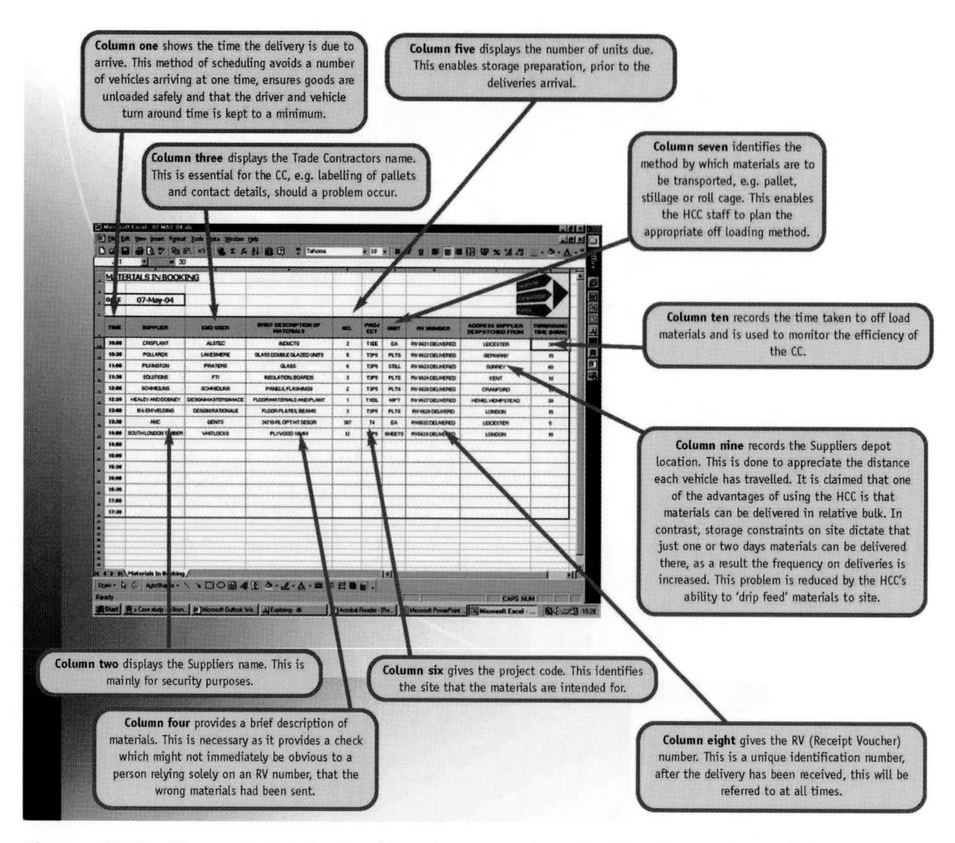

Figure 11.2 The materials-in booking form used at the Heathrow Consolidation Centre

least 24 hours prior to delivery to enable efficient resource management and to reduce potential bottle-necks. The information required includes:

- trade contractor's name
- details of the carrier/haulage company delivering the consignment
- project code
- intended date of delivery
- details of goods, i.e. number of pallets and a brief description
- anticipated duration of stay at the CC.

Figure 11.3 depicts the 'high level process map' demonstrating the flow of materials through the HCC, from the initial identification by the trade contractor to the JIT delivery of the materials to the pre-agreed workface location.

Goods-in process: offloading and checking procedures

Upon arrival at the HCC, a logistics operative will inspect the consignment, checking that there is no damage and that the pallet contents

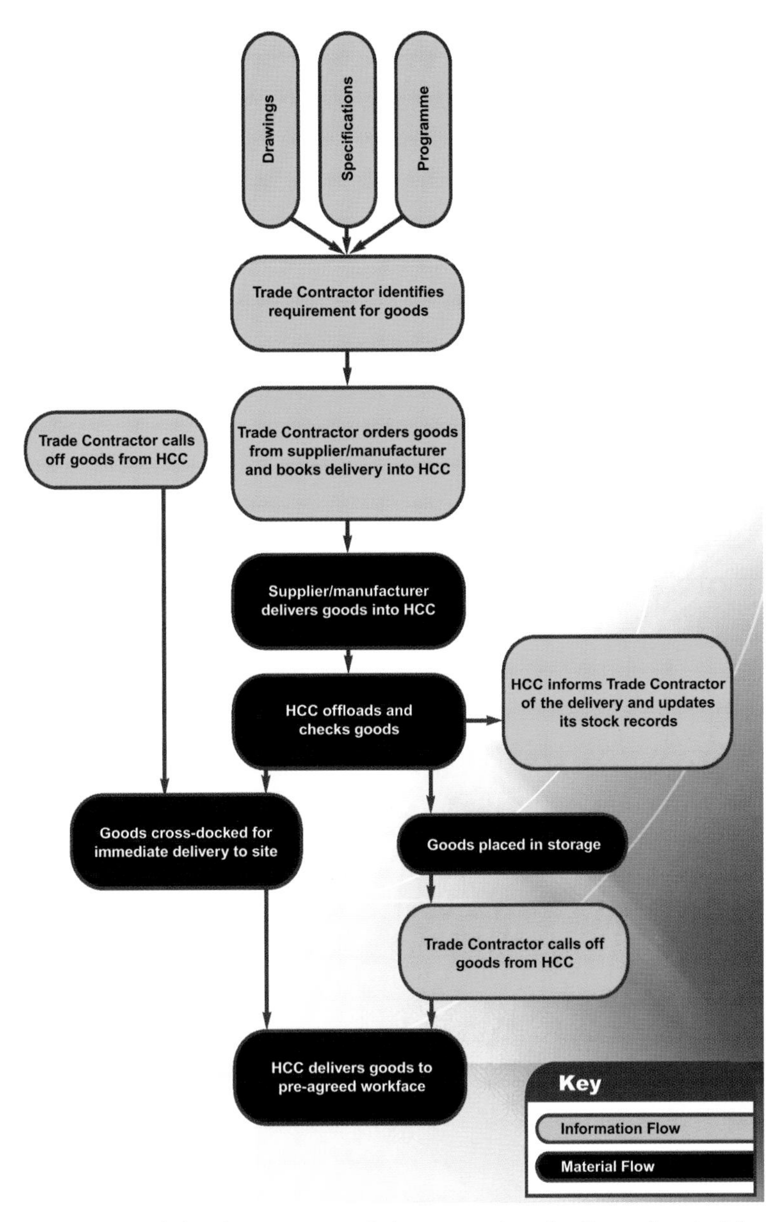

Figure 11.3 'High level process map' demonstrating the flow of materials through the Heathrow Consolidation Centre

correlate with the delivery note. Goods are not removed from the pallet and packages are not opened, first because the transit packaging is still required to complete the final leg of the delivery to site and, second, because logistics operatives do not normally have sufficient knowledge of all material types to enable them to detect a problem (unless it is very obvious), due to the extensive range of materials handled.

Figure 11.4 An accepted and labelled pallet

If a discrepancy with a delivery is discovered, the CC contacts the trade contractor immediately and a decision is made whether or not the delivery should be accepted. If rejected, no further action is required by the CC because the responsibility for organising replacement materials lies with the trade contractor. If accepted, a logistics operative will authorise and sign the delivery note and label the pallet(s) (Figure 11.4) before placing them onto the pallet racking, bulk storage or 'cross-docking' for immediate dispatch to site. Displaying this information on a label enables the CC staff to quickly assess the contents of a pallet.

Goods-in process: storage

The storage location of each consignment may be predetermined, depending on the sophistication of the WMS. Using the most basic paper-based system, the storage location is subject to space availability. However, a specialist WMS would be capable of allocating each pallet to a suitable space in the warehouse as soon as the data is placed on the system (e.g. when the materials are booked in by the trade contractor) 24 hours before the delivery arrives.

At the HCC, each pallet space is assigned a unique location reference. The location of the materials is recorded on a stores layout, a delivery note and the 'customers' stock held register', an inventory record. This enables the CC staff to locate materials when preparing for a future delivery to site and enables the supply controller (the person who has

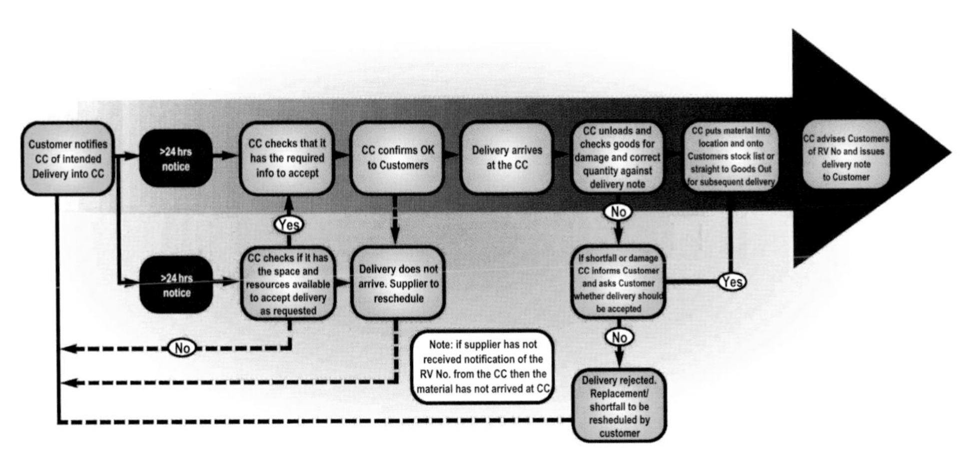

Figure 11.5 The procedures followed when materials are delivered into the Heathrow Consolidation Centre

responsibility for managing trade contractor accounts) to monitor material stock levels.

In the event of materials being damaged as a result of poor handling at the CC, the manager will investigate the cause and report the incident to the trade contractor. The cost of replacement (if required), subject to insurance conditions, is funded by the HCC. Figure 11.5 illustrates the procedures followed when materials are delivered into the HCC.

Goods-in process: trade contractors' inventory records

After materials are placed in storage, the delivery note is passed to a stock controller who will update the trade contractor's inventory records. Figure 11.6 demonstrates the form developed by the HCC for this purpose.

Goods-out process

The procedure leading to materials being delivered to the site workface begins with the trade contractor submitting a 'call off request' (Figure 11.7). This document can be transmitted by fax, email or post. Verbal requests from the trade contractor, although more expedient, are not considered appropriate as vital information might be excluded, and no audit trail is available if a query arises. The information collated on the 'call off request' includes:

- the trade contractor's name
- a contact name and number

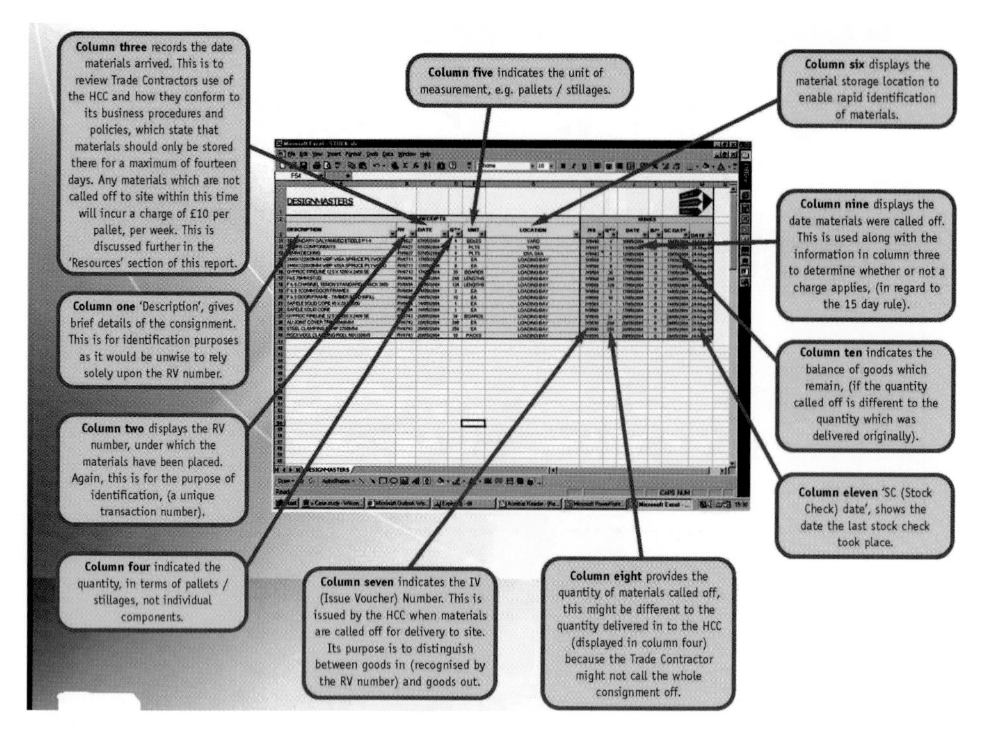

Figure 11.6 A trade contractor's inventory record form used at the Heathrow Consolidation Centre

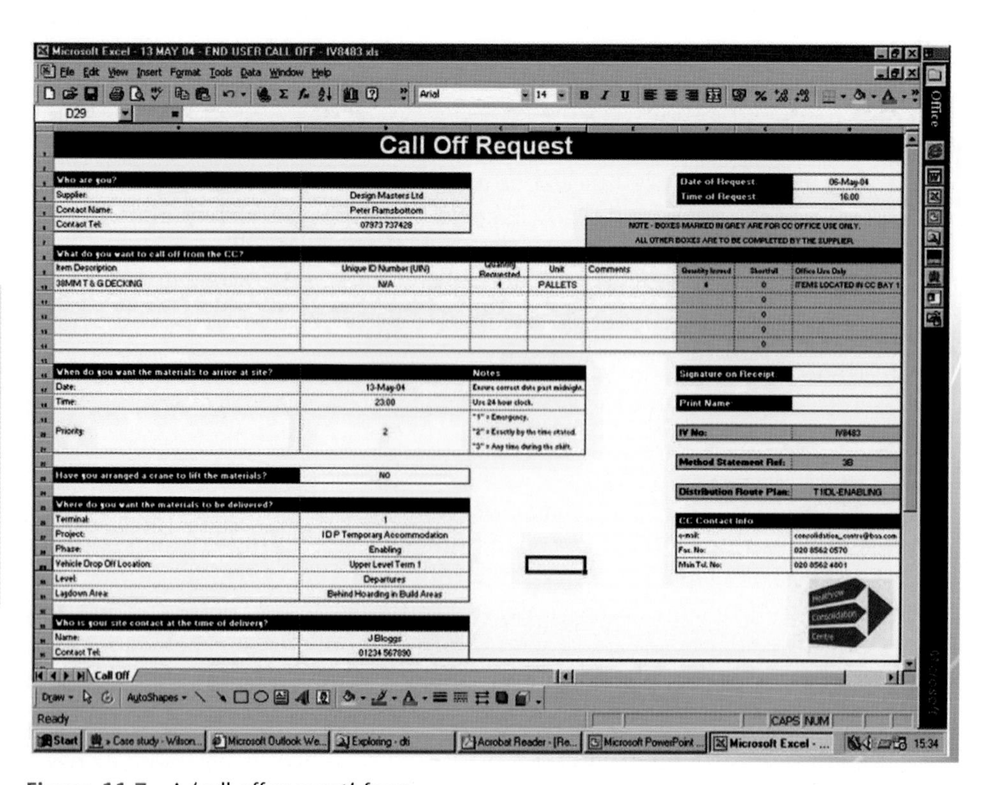

Microsoft Excel - 13 MAY 04 - END USER CALL OFF - IV8483.xls

Call Off Request

Who are you?	
Supplier:	Design Masters Ltd
Contact Name:	Peter Ramsbottom
Contact Tel:	07973 737428

Date of Request	06-May-04
Time of Request	16.00

NOTE - BOXES MARKED IN GREY ARE FOR CC OFFICE USE ONLY.
ALL OTHER BOXES ARE TO BE COMPLETED BY THE SUPPLIER.

What do you want to call off from the CC?

Item Description	Unique ID Number (UIN)	Quantity Requested	Unit	Comments	Quantity Issued	Shortfall	Office Use Only
38MM T & G DECKING	N/A	4	PALLETS		4	0	ITEMS LOCATED IN CC BAY 1
						0	
						0	
						0	
						0	

When do you want the materials to arrive at site?		Notes
Date:	13-May-04	Ensure correct date past midnight.
Time:	23.00	Use 24 hour clock.
Priority:	2	"1" = Emergency. "2" = Exactly by the time stated. "3" = Any time during the shift.

Have you arranged a crane to lift the materials?	NO

Where do you want the materials to be delivered?	
Terminal:	1
Project:	IDP Temporary Accommodation
Phase:	Enabling
Vehicle Drop Off Location:	Upper Level Term 1
Level:	Departures
Laydown Area:	Behind Hoarding in Build Areas

Who is your site contact at the time of delivery?	
Name:	J Bloggs
Contact Tel:	01234 567890

Signature on Receipt	
Print Name	
IV No:	IV8483
Method Statement Ref:	3B
Distribution Route Plan:	T1IDL-ENABLING

CC Contact Info	
e-mail:	consolidation_centre@baa.com
Fax. No:	020 8562 0570
Main Tel. No:	020 8562 4801

Heathrow Consolidation Centre

Call Off

Figure 11.7 A 'call off request' form

- description of materials
- unique identification number (UIN)
- quantity of materials
- delivery location on site
- a contact name and number of a person who will be on site at the time of a delivery (this may differ from the contact details given above, because of shift work).

A record is made of the time and date the call off request is received to determine the length of time materials have been stored at the HCC.

It is important that when submitting the call off request the trade contractor provides sufficient time to allow the CC to plan resources, prepare the materials for delivery and, in exceptional cases, hire additional mechanical resources when, for example, the current level of resources is not sufficient to meet the demand or where a consignment has unusual lifting requirements. Initially, the HCC demanded 48 hours' notice; however, many trade contractors stated that this was *often difficult to plan that far ahead*. It was also apparent that this period of time was rarely needed by the logistics operatives to check that sufficient stocks were available to satisfy call off requests. In response to this, the HCC later reduced the 48-hour notice period to 24 hours.

A daily work pack summary document is generated for each call off and is used to prepare the materials. In some cases, where the materials are easily recognised, for example bricks, blocks and plasterboard, the HCC is able to provide the trade contractor with the precise quantities required to fulfil a task on site, and these are known as 'work packs'. The extent to which this can be achieved is, however, limited by the lack of specialist materials knowledge of the HCC operatives and the vast range of materials handled.

A supervisor reviews the contents of the work pack summary and distribution report and considers the following:

- the nature of the consignment, e.g. special handling requirements
- the logistical difficulties in executing the delivery, e.g. vehicle access constraints
- additional safety precautions, e.g. road closure, safety barriers and signage
- the sequence of delivery drops that optimises the most efficient routes.

Delivery preparation takes place hours before the delivery is due on site. After reviewing the deliveries for the day, the HCC supervisor issues 'pick sheets' to the logistics operatives, who select the materials from the pallet racking and place them in the goods-out bay. Materials remain in the goods-out bay until loaded, just minutes before the delivery vehicle departs.

As part of this process the CC manager briefs the site-based supervisor with delivery-specific details, for example special handling requirements, the expected time of arrival and the necessary method statements, risk assessments, lifting plans and Control of Substances Hazardous to Health (COSHH) assessments. This also gives the opportunity for the supervisor to advise the HCC about any difficulties which may be encountered en route to site, for example congested areas or roadworks planned or in progress.

After the delivery is received on site, the signed delivery note is returned to the CC and the trade contractor's account (inventory levels) is adjusted on the stock held register. It is important that this information is kept up to date and that each trade contractor has access to this information at all times.

A significant benefit of this methodical data entry communication process is that a robust audit trail of all transactions is produced. Should a query arise concerning the timing or location of a consignment, supporting information is readily available.

Warehouse management systems

The principal reasons for using a WMS are:

- The importance of creating a reliable means of processing deliveries. A trade contractor's productivity relies upon the punctual availability of materials. This factor ultimately affects overall project reliability. It is important to create a robust audit trail to ensure that each consignment can be traced throughout each stage of delivery.
- The need to create a standard means of product description between the logistics team at the CC and trade contractor on site is essential. The HCC began by using a basic paper-based system. Initially, this was developed to overcome the problem caused by different parties referring to the same product by different names. This system was later adapted to perform more advanced applications, such as the production of graphs to monitor the performance of the delivery team.

WMS are commonly used in distribution centre environments to reduce the burden of data entry and to provide an efficient means of controlling stock inventory. Their functionality varies from a basic system capable of recording the flow of materials through the warehouse to advanced systems able to consider optimum storage locations based on the shape and size of a consignment. For example, an advanced WMS might be capable of volumetric calculation or ABC storage. The former is where the software is able to record a continuous, real-time picture of the free space available in the warehouse and its location and to compute instantly and match this with scheduled 'goods in' that considers height, width and depth – giving an instant 'put away decision'. The latter is able to allocate storage zones based on the materials' frequency of use – the most commonly used materials are located in prime areas of the warehouse, for example nearest to the access doors of the warehouse (location A). Similarly, goods which are picked less frequently would be located in aisles designated 'B', and those with the slowest rate of turn over, designated 'C', are located furthest from the door. Additional features include links to invoicing packages, payment modules and auto pick-list creation.

The level of success demonstrated at the HCC proves that such an advanced WMS is not required. A simple spreadsheet is adequate. However, as the concept develops and is applied to different situations, operational demands may require a greater level of sophistication, such as the need to control storage time.

A further example of a more advanced WMS includes the forward matching of supply and demand. Matching supply with demand is usually only possible once goods have been received at the distribution centre. In a number of cases, the HCC scheduled the distribution of goods before confirmation that the materials had previously been delivered. However, without upstream supply chain visibility this process is capable of providing last-minute surprises and complications. Fulfilment planning has the ability to match advance shipment notice (ASN) or even pending procurement orders to anticipated outbound flow. It features the ability to instruct suppliers of the exact arrival time at the CC. This information is derived from the time of the actual order, the lead time and considers all bottlenecks. As a result, a pull-based cross-dock process can be established for suitable suppliers and their products. This enables the CC to cope with line proliferation and process higher product volumes and react to overloads before bottlenecks arise.

Figure 11.8 illustrates the 'expense' versus 'functionality' range when considering a WMS.

Materials identification and tagging systems

The range and volume of materials processed by a CC are such that it requires a management system capable of monitoring and controlling stock levels and locating and facilitating the identification of components. This chapter considers the application of various tools which achieve this and promote efficient logistics by improving communication of product information throughout the supply chain. There are essentially three options, with varying levels of sophistication available:

- a paper-based system
- barcode technology
- RFID systems.

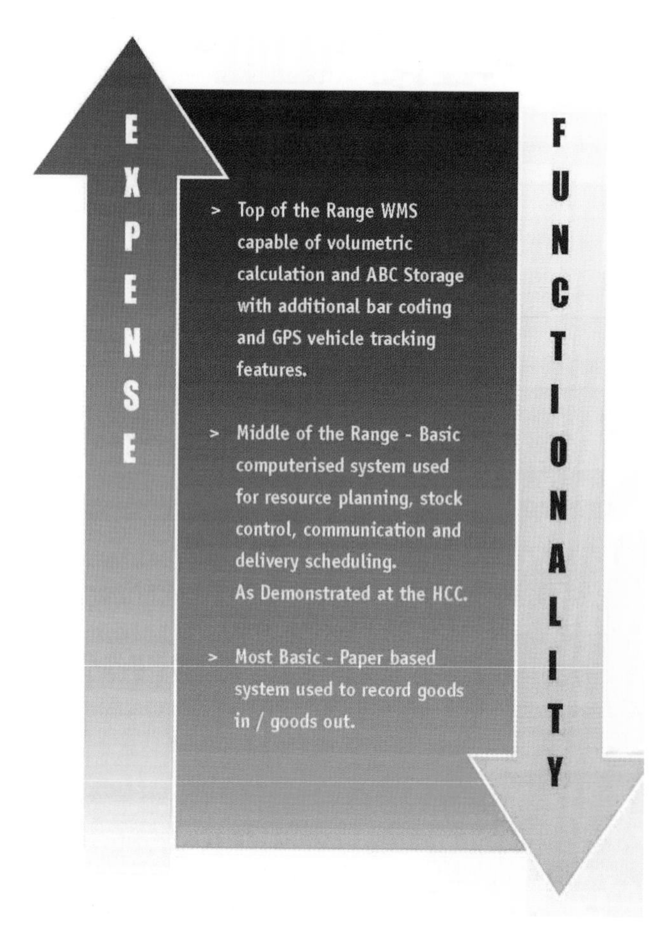

Figure 11.8 The 'expense' versus 'functionality' range when considering a warehouse management system

The information needed to achieve this requires project details, material type, quantity, owner details (e.g. the trade contractor), delivery details such as the time and location the materials are required on site and any health and safety data such as COSHH data sheets which must accompany the delivery.

Paper-based system

Using a paper-based system, each consignment will be allocated a UIN. The UIN (Figure 11.9) is displayed on the materials and the corresponding information is stored on a spreadsheet. This is an inexpensive and effective method of controlling the flow of materials.

The HCC began by using a paper-based system. To collate the necessary information, each trade contractor was asked to create a 'catalogue'

Product Code	Description
TC001	FOCUS 'D' CEILING TILES (1800 x 600 x 200)
TC002	GYPSUM JOINT CEMENT
TC003	SAS Sys. 130 575 x 575MM Perf 1522
TC004	SAS 600 x 600 x 40 MM TRUCELL T15 CEILING PANEL
TC005	KNAUF COLUMN FIXING CLIPS
TC006	PRELUDE CROSS TEE BARS 600 MM
TC007	GYPSUM PAPER JOINT TAPE x 150 MM
TC008	GYPSUM MF9 CONNECTOR CLIPS
TC009	SELF DRILLING SCREWS 25 MM
TC010	CROSS TEE BARS 1200 MM
TC011	SAS SYSTEM 130 (GREY) TILES 1200 x 600 MM
TC012	STEEL DIMOND MESH 2.5 x 1.2 x 2 MM

Figure 11.9 Example of material product unique identification numbers

of their most commonly used products. These typically consisted of the details of up to 60 items, which were then given a UIN. After this point, products were described principally by their UIN (with a brief description purely as a check). This technique eliminated any confusion caused by different members of the same trade contractor organisation describing products by a variety of terms. As the products used by a trade contractor are subject to constant change throughout the duration of a project, the development of these catalogues is ongoing.

Barcoding and radio-frequency identification technology

Barcode technology has been widely adopted by the retail, defence and manufacturing industries. Even though its use has produced substantial savings and it has proved an efficient and accurate method of recording data, the barcode offers limited data storage capacity and data transfer from the barcode depends on a direct line of sight with the reader. A more advanced version of the barcode is the RFID tag. The main advantages of RFID technology is that it is capable of storing larger volumes of information, and data transfer does not require direct line of sight between the tag and the reader.

A lack of collaboration between organisations has hindered the growth of such technologies, as the effective application and utilisation of barcoding and RFID technology depends on the use of a standardised programme throughout the supply chain. This section considers the application of barcode and RFID technology to the construction supply chain.

Barcoding

A barcode is a machine-readable symbol used to encode product information such as its UIN and cost. The data contained within the barcode is transferred to a reader when scanned. Barcodes are commonly used to control stock in a warehouse and are normally used to transfer product information throughout the supply chain.

A trial conducted at Heathrow Airport examined the potential benefits of using barcoding in a construction environment. The trial considered the flow of materials from the supplier to delivery on site and found that its application provided the following benefits:

- increased pick accuracy (from 98.85% to 99.6%)
- increased speed and certainty when validating the location of delivered goods on site
- created a robust audit trail
- provided a rapid and accurate means of checking stock
- reduced worker hours, as stock changes are automatically uploaded to reflect inventory levels.

However, the trials also revealed that few manufacturers were either prepared to or capable of applying barcode labelling to materials before they left their stores. This meant that labels were attached as deliveries arrived, which caused delay in the receipt and storage of materials. It is important to note that the benefits of barcode technology can only be maximised if used in conjunction with a complementary 'back end' application, such as a WMS.

Figure 11.10 shows the handheld, barcoding device used at the HCC.

Figure 11.11 shows the 'goods out' process used at the HCC.

Figure 11.12 indicates the way in which barcoding technology could be incorporated into the current process of materials delivery.

Radio-frequency identification technology

RFID technology (also known as 'e-tagging') is an automated data transfer between a tag and a reader which provides an accurate method of

Figure 11.10 A handheld, barcoding device used at the Heathrow Consolidation Centre

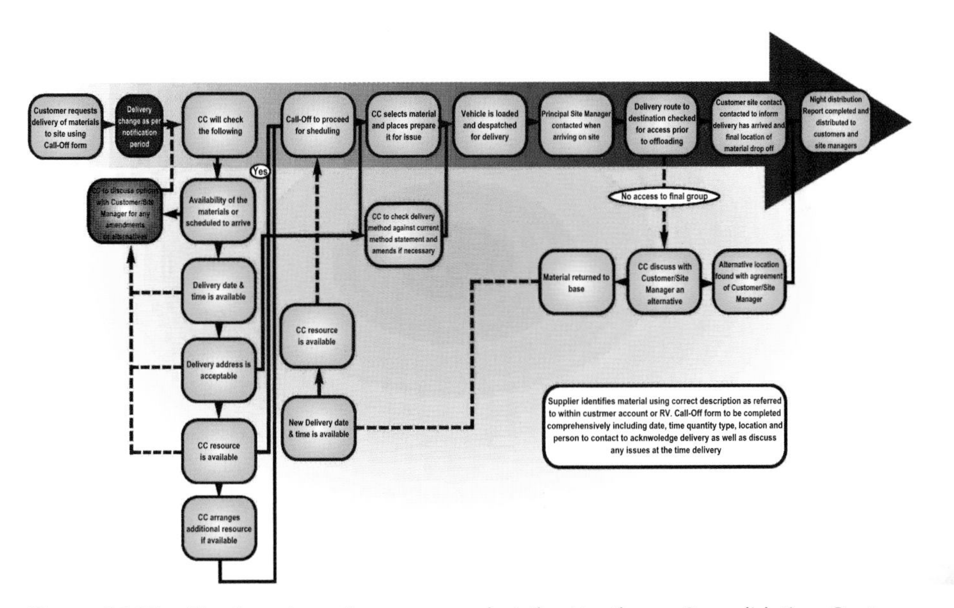

Figure 11.11 The 'goods out' process used at the Heathrow Consolidation Centre

tracking materials, controlling inventory and improving communication throughout the supply chain (Collins 2004). Powered by an inbuilt energy source, the tag or RFID 'transponder' consists of a microchip and antenna. The information contained on the microchip is transmitted to

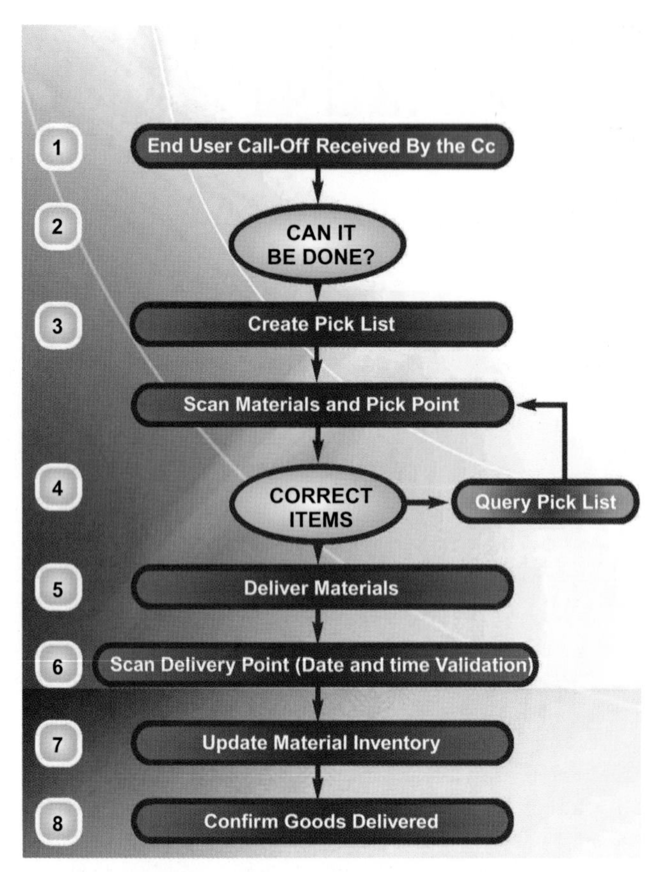

Figure 11.12 How barcoding technology could be incorporated into the current process of material delivery

a reader that converts the radio signal into a digital signal which can be read and communicated to a data network. RFID tagging is a more sophisticated IT solution than barcoding for a number of reasons:

- Tags are capable of storing more information than the standard barcode.
- The technology is capable of reading through most materials. Therefore, tags can be attached to a material e.g. embedded into a component or pallet or temporarily fixed to the outside and removed (if necessary) once incorporated into a buildings fabric, Collins (2004);
- Information can be obtained from multiple tags at one time whereas barcode technology depends on direct line of sight and is only capable of obtaining information from one source at a time. With a read range of up to 3 m, this allows tags to be identified from the back of a lorry (providing that the vehicle stops momentarily).

- The technology can enhance a building's post-construction performance, e.g. by providing information on service records of mechanical and electrical plant or warning when a structural member is overloaded (Collins 2004).
- Data contained on the tag can be updated and rewritten as it travels through the supply chain. Its route and status can be recorded to assist with and facilitate VSM.

Value stream mapping

VSM originated in the car manufacturing industry as a tool for implementing lean production (Rother and Shook 2003). Also known as 'value chain analysis', it is an analytical process for understanding an organisation's methods based on individual activities to ultimately determine the value added at each stage (Bocij *et al.* 2003).

VSM aims to eradicate waste from the production process by identifying non-value-adding elements, as determined by the client's perception of 'value'. To achieve this, both the information and materials flows of a single product and its components are mapped from start to finish, this includes the design flow from concept to handover and the production flow from raw materials to completion (Rother and Shook 2003), providing a clear picture of how effectively resources are being used throughout the supply chain (Bocij *et al.* 2003).

Figure 11.13 illustrates a typical supply chain and considers the flow directions of materials and information.

The basic mapping procedure initially involves drawing a visual representation of each process involving materials and information from the client to the supplier. Value stream mapping involves identifying and documenting the processing steps for each product family. Following this it is necessary to consider the value input of each section before producing a 'future state' map, which should exclude any section which does not contribute adequately to the value of the final product. Figure 11.14 illustrates the initial VSM steps.

To analyse each organisation's contribution to the supply chain, it is necessary to distinguish between 'primary activities' and 'support activities' (Bocij *et al.* 2003).

Primary activities are activities that contribute directly to getting goods and services closer to the customer and include:

- inbound logistics
- internal activities, e.g. the production process
- outbound logistics
- marketing and sales
- after-sales service.

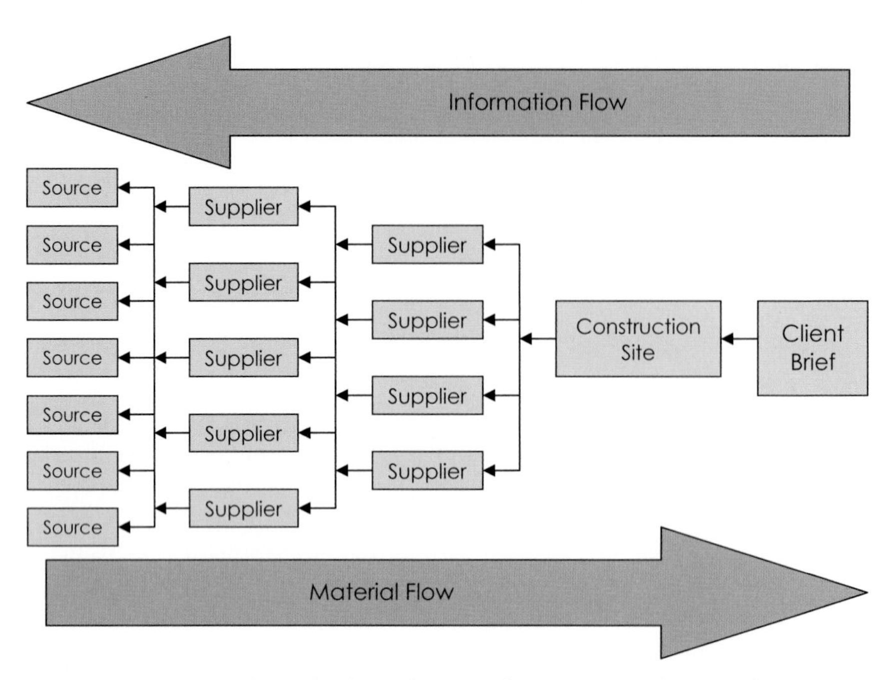

Figure 11.13 Typical supply chain direction flow of materials and information

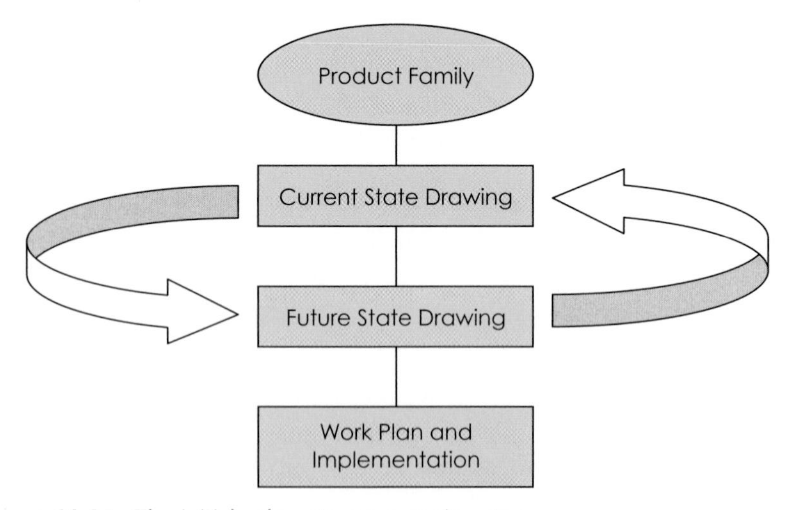

Figure 11.14 The initial value stream mapping steps

Support activities provide the inputs and infrastructure that allow the primary activities to take place. These include:

- corporate administration and infrastructure
- human resource management
- technological development
- procurement.

Comparing value stream mapping in the manufacturing and construction industries

The retail and manufacturing industries use VSM to refine the flow of materials from the source of origin to the end user. This has been done to good effect as consumers demand better-quality products at lower prices. However, analysis of the value stream in construction is commonly limited to the final link of the supply chain, for example the supplier/merchant to the construction site, and is often as basic as selecting the supplier that provides the lowest price.

The bespoke design of a building, the specification of non-standardised products and the relatively short duration of most construction projects make it difficult to track a value stream, because of the number and variety of components used. However, even when there is a high level of product certainty and predictability over a sustained period, for example the construction of a standard house type, little effort is made and little is known about the value input of each element of the supply chain.

The complexity of the process and number of parties involved also hinders knowledge of the supply chain. Johnson and Scholes (1999) state, 'It is very rare that a single organisation undertakes all of the value activities from the product design through to the delivery of the final product or service to the final consumer.' This is particularly true of a construction project with different consultants responsible for:

- establishing the design criteria for the building
- determining ground conditions of the site
- designing elements of the building
- designing components of the building
- supplying materials
- carrying out work packages on site
- managing the construction work on site.

Being able to differentiate between inbound logistics, the production process and outbound logistics is a capacity that is often lacking by those stakeholders involved in the construction supply chain. This makes it difficult for the end user to understand and breakdown the costs of materials. Suppliers who claim to provide a 'free' delivery service add to this problem as their costs are inevitably integrated with the cost of the product.

Figure 11.15 illustrates that there are a number of factors that have financial consequences in the traditional supply chain for construction materials and demonstrates the difficulties encountered when pricing logistics or analysing the value contribution of each activity. For example, two issues make it difficult to measure the cost of logistics: the nature

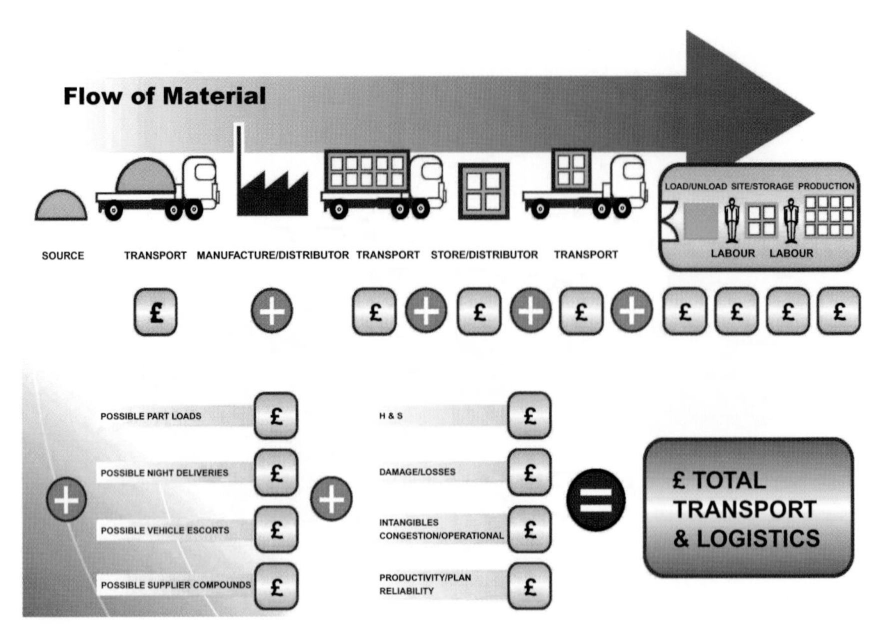

Figure 11.15 Flow of material

of factors such as health and safety and their effect on productivity and, second, the number of delivery variables such as part loads and multi-drop deliveries.

When compared with the manufacture of a car, which has a known product base that places a high level of certainty on the number of components used, it is clear that value mapping principles are far easier to implement. To complement this, the design process carefully considers production techniques to ensure optimum efficiency. The controlled factory environment also makes it easier to analyse the value contribution of each input as each person and section has a clearly defined role to fulfil.

Figure 11.16 summarises the factors which facilitate VSM between the construction and manufacturing industries.

This reactive approach provides significant potential to improve the quality of construction products and reduce cost. Bespoke products and non-standardised components are frequently used in construction projects, whereas in retail and manufacturing there is more scope for standardisation, VSM and supply chain analysis. However, even in the sectors of the construction industry that have a high potential to benefit from VSM and supply chain analyses, such as volume house building, little evidence exits of their adoption or application. The CCM, however, successfully demonstrates an application of supply chain integration and value mapping processes.

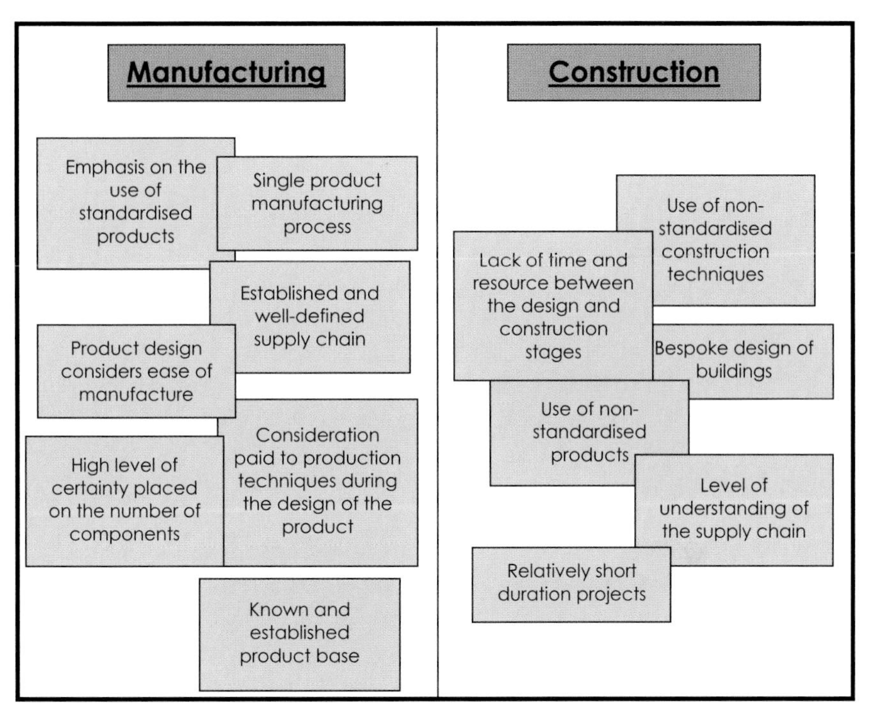

Figure 11.16 Factors which facilitate value stream mapping between the construction and manufacturing industries

The cost benefits of a consolidation centre

As well as providing significant advantages to a contractor, it is probable that other parties including trade contractors and materials suppliers benefit from using a CC; however, their cost savings are difficult to quantify. Despite considerable effort at Heathrow, two factors have prevented cost savings being passed to the client. First, few in the construction industry fully understand their logistics costs. This lack of understanding is obstructing savings being passed on to the client. Second, trade contractors are reluctant to share financial data for fear that the client will seek to reclaim any savings made.

Trade contractors

During the study carried out at Heathrow Airport, it was found that two out of every three trade contractors, agreed that the HCC had brought cost savings to their organisation. The exact level of their savings, however, was difficult to quantify.

The various ways in which a trade contractor benefits financially as a result of a CC include:

- the ability to arrange deliveries in relative bulk (even when there is inadequate storage space available on site)
- the reduced need for site-based labourers
- less damage to materials
- improved health and safety standards
- better use of the trade contractor's skilled trades workforce.

Calculating the extent of these savings is difficult because of their dependence upon a variety of circumstances. For example, whilst it is reasonable to expect that cost savings can be made by purchasing and delivering materials in bulk it is difficult to quantify the specific savings made compared to alternative delivery options.

Analysing the way in which time is spent on site

Cost savings are made as a result of better utilisation of the skilled trades workforce. Figure 11.17 illustrates the various causes for unproductive time on site. The productivity of operatives on site demands particular attention especially during periods of skills shortages in the industry.

An investigation conducted by Lloydmasters (www.lloydmasters.co.uk) on behalf of Wilson James gave an insight into the amount of time which is wasted on site. The investigation examined the amount of time the skilled workforce spent carrying out various tasks on five projects which were dependent upon traditional logistics techniques. The results are displayed in Figure 11.18. The only truly productive option in terms of value-adding tasks demonstrated is 'construction activity',

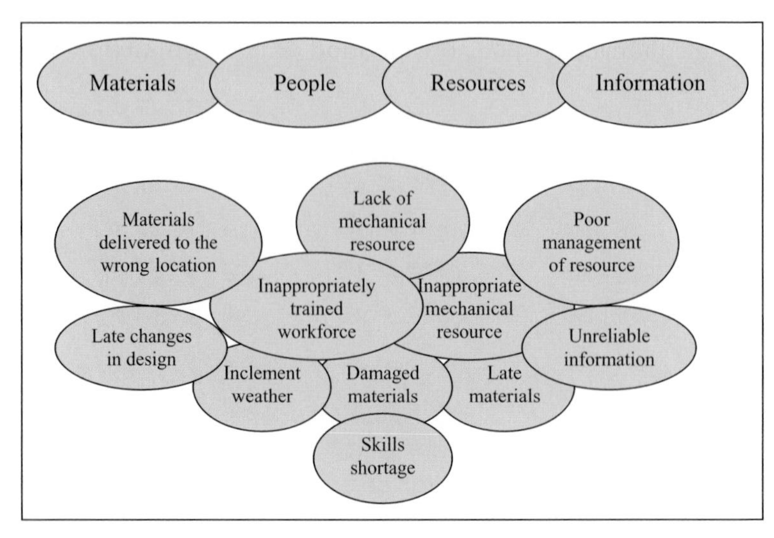

Figure 11.17 The various causes for unproductive time on site

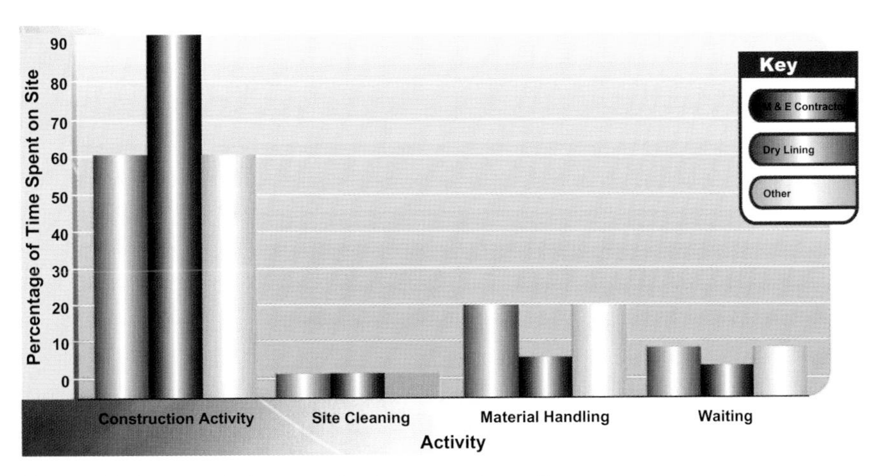

Figure 11.18 Breakdown of the time spent on site

whereas 'site cleaning' and 'materials handling' are only semi-productive as they could have been carried out by less qualified personnel. This misuse of resources suggests that project performance could be enhanced by clients considering a holistic logistics service which promotes time certainty by eradicating the misuse of the skilled workforce's time. However, it cannot be assumed that this alone will achieve maximum output, as other factors, such as an individual's organisation, attendance, punctuality and work ethic, need to be considered. For instance, Lloydmasters' report fails to identify 'idle time' and as a result the study assumes that operatives are actively participating in one of the options for the entire time they are in work. Furthermore, the study suggests that any 'waiting' time is spent waiting for materials. However, it is possible that the equivalent amount of time would be lost by some of the individual's poor organisation or poor work ethic, which is present on many sites.

Based on a tradesperson working a six-day week with four weeks' annual leave on a typical 12-month construction project suggests that up to 100.8 days would be spent unproductively. These would consist of 8.64 days 'site cleaning', 20.16 days 'waiting' and 72 days 'materials handling'. In an attempt to convert this time into a cost, the above percentages of time spent on site were divided into the notional cost of a project. For example, on a project worth £25 million, £11 million is spent on labour, £220,000 of this is spent cleaning, £770,000 is spent waiting for materials, £1,430,000 is spent handling materials and just £8,580,000 (34%) of this funds 'construction activity'.

Significantly, the Lloydmasters' report acknowledges that 'much of the non-construction cost is hidden; it is not known or measured by either the trade contractors or main contractor'. This confirms what had been thought previously, that the accounting systems practised within the construction industry do not facilitate the precise itemisation of logistics costs. One possible explanation for this is the extent to which the means of getting materials to site can vary, for example a simple change in design or delay on site could mean that materials are not used in the sequence that was originally envisaged and as a result get moved around more. Such events would therefore be extremely difficult to cost.

Increasing the availability of materials

Figure 11.19 compares the number of tasks completed successfully before November 2001 (pre-CC) with those tasks successfully completed after November 2001 (post-CC). The results indicate that the HCC reduced the number of delays due to materials not being available from 6% to 0.4%, representing a reduction factor of 15. This increase in certainty reduced the associated cost of delays on site, rescheduling of work and ultimately the risk of over-running project deadlines. Interestingly, there was a 11% overall improvement after the HCC was implemented as tasks incomplete due to 'other' reasons also fell; however, there is no evidence to suggest that this is a direct result of the HCC.

However, the CC's ability to improve plan reliability on site depends upon the:

- trade contractor ordering the materials on time
- materials supplier delivering the correct materials on time
- materials being in good condition when they arrive at the HCC.

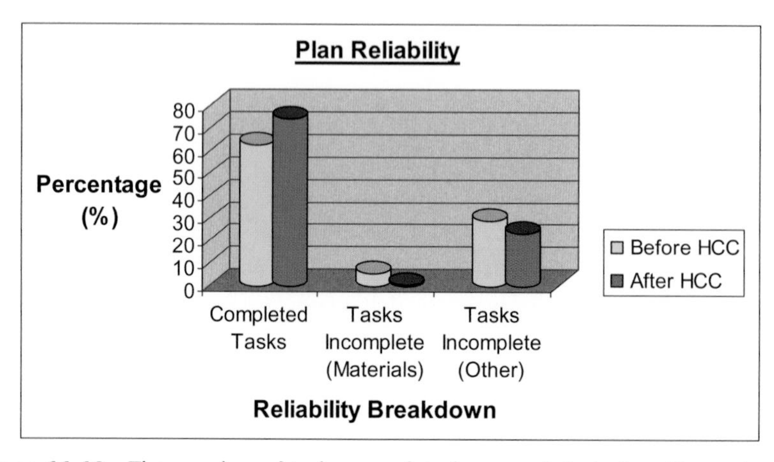

Figure 11.19 The number of tasks completed successfully before November 2001 (pre-consolidation centre) with those tasks successfully completed after November 2001 (post-consolidation centre)

Advantages offered to materials suppliers: better use of delivery vehicles

To explore the benefits that the CCM offers to materials suppliers, Figure 11.20 compares the speed of offload at the CC, a site in central London and a site at Heathrow and demonstrates that the deliveries received by the HCC reduce driver turn-around time and increase delivery time predictability for suppliers. It demonstrates that for every one delivery received airside at Heathrow 3.65 deliveries could be received at the HCC and for every one received at a site in central London 5.73 deliveries could be received at the HCC. Potentially, materials suppliers could reduce their transportation costs by up to two-thirds if the performance consistently achieved by the HCC represented the industry norm.

The environment provided by the CC is geared towards delivering efficient logistics. The HCC is solely dedicated to the purpose of handling, storing and delivering construction materials and equipment. Deliveries are scheduled to avoid a number of vehicles turning up at one time. The unloading process starts as soon as the delivery vehicle arrives and is met by a team capable of offloading/loading deliveries with a variety of mechanical aids, such as forklift trucks. This is very different from the majority of construction sites, where there is less space to

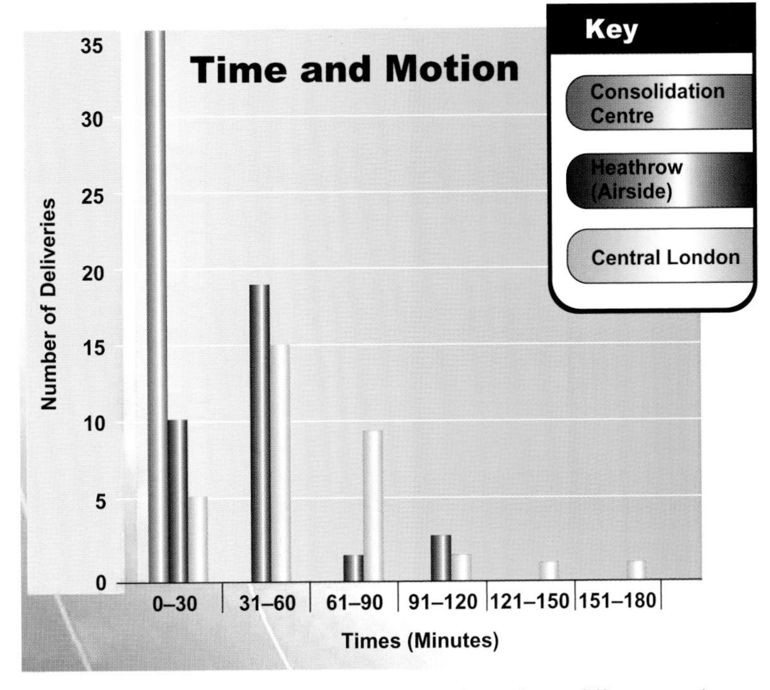

Figure 11.20 Comparison of offloading speeds at three different projects

manoeuvre vehicles, delivery bays are exposed and there is normally just one person capable of driving the only forklift truck on site. Furthermore, the processing of goods on site is at times chaotic, which owes much to a less robust delivery schedule. Even where scheduling exists, congested road conditions make it much less certain that the consignment will arrive on time.

Increasing driver and vehicle productivity by offering increased offload time predictability enables drivers to complete more drops per day. This not only reduces the cost of each delivery but also reduces the restrictions placed on drivers by the Working Time Regulations (1998), which implement the European Working Time Directive and stipulates that, an 'adult worker is entitled to a rest period of not less than eleven consecutive hours in each 24-hour period which he works for his employer' (HMSO 1998).

The costs incurred by a materials supplier while delivering materials are associated with the cost of a delivery vehicle and driver. Considering the following elements, a 17.5 tonne rigid-bed lorry would typically cost around £40 per hour based upon the following assumptions:

- vehicle purchase/hire cost (£750 per week)
- insurance, tax, maintenance (combined £250 per week)
- driver-associated costs (£1000 per week)

Fuel costs were not taken into account since the distance travelled by each vehicle on each delivery is likely to vary. When applying this cost to the average times taken to deliver to the three areas studied, it becomes clear that suppliers and trade contractors are making a cost saving.

Table 11.1 compares the cost of delivering to the HCC, a central London project and a Heathrow airside project based on a time and motion survey conducted for the DTI (Department of Trade and Industry 2004a).

Figure 11.21 demonstrates that, based on offload times alone, it is almost six times more expensive for suppliers to deliver directly to site in central London and over 3.5 times more expensive to deliver directly to a project located airside at Heathrow than it would be to the HCC.

Table 11.1 Time/cost comparison per delivery.

Time/Cost Comparison Per Delivery		
	Average time (min./sec.)	**Cost (£)**
HCC	11.26	7.62
Central London	64.56	43.29
Heathrow (airside)	41.15	27.50

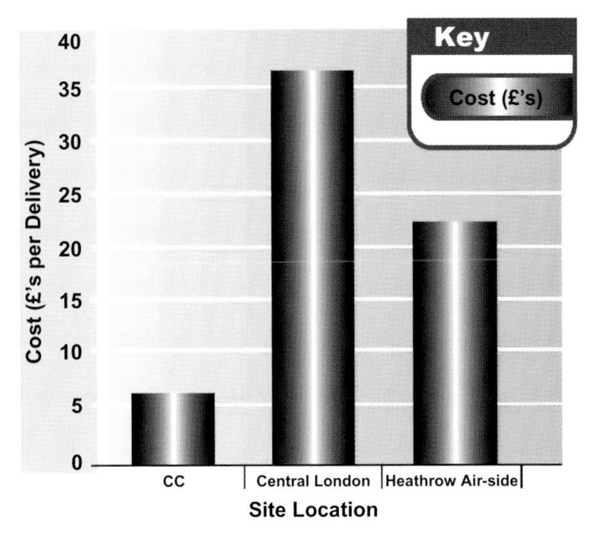

Figure 11.21 Comparison of delivery cost to suppliers at three different projects

Considering that one delivery vehicle may make up to six deliveries per day, the potential savings for suppliers and trade contractors is considerable. The amount of the potential savings varies depending on the location of the site.

Realising cost savings

Two issues make it difficult to demonstrate the cost savings associated with a CC: an inability to recognise the potential supply chain benefits and the imprecise method that organisations use to fund their logistics costs. Naturally, the supply chain is reluctant to surrender cost savings as this reduces their profit margins. This has been a reoccurring problem for making cost model comparisons with a CC.

Where CCs are in operation, trade contractors have been used as a cost buffer. The client consistently pursues the savings made as a result of the efficiency gains provided by the CC but the trade contractor is unable to reclaim these costs from their materials suppliers. To solve this, the method by which materials are priced needs to be reformed and refined throughout the supply chain to increase visibility. Materials suppliers must be open about their logistics costs and a more scientific approach must be adopted for pricing items. Their storage, handling and transportation costs must be analysed and added to the factory gate price of the product. This information must therefore be available to other parties further down the supply chain in order to eradicate non-value-added costs.

Cost savings resulting from the use of a CC are present in various forms, although the nature of these savings makes it difficult to quantify. Claims by many materials suppliers that delivery is 'free' are unfounded and add confusion to the industry's already limited understanding of logistics costs. The various stakeholders throughout the supply chain must recognise and acknowledge their logistics costs and make this information available to others if the cost benefits are to be fully realised. Cost itemisation must therefore include the factory gate price of the product and the logistics costs to increase cost visibility throughout the supply chain.

Non-financial benefits of the consolidation centre

This section considers the non-financial benefits arising from the use of a CC, for example accountability, security and environment.

The initial driver for the HCC was to 'deliver major construction projects in accordance with BAA's construction impact strategy' (Chris Ctori, T3 Project Team Leader, in interview with the author, February 2003). This stipulated the need for a harmonious integration of construction work with the client's primary business objective, which included the efficient service delivery to passengers. BAA's 'construction impact strategy' clearly indicated that a more accountable system of materials delivery was required. It was soon apparent that the HCC achieved this objective by providing an integrated logistics service. However, there were other advantages of having a specialist contractor completing the final leg of the materials journey to site. These were mainly correlated to reduced vehicle movements and are intrinsically linked to the CC's ability to consolidate numerous consignments to one vehicle for delivery to site.

Increased accountability

Relying on traditional logistics techniques, with each trade contractor having their own fragmented supply chain, can cause major disruption to the community immediately adjacent to a construction project. Driver surveys conducted in central London as part of the Department of Trade and Industry's final report 'Consolidating Construction Materials Transferring the Methodology' (2004b) revealed that the vast majority of materials suppliers meet scheduled delivery slots by arriving up to one hour early and either parking outside the site or driving around adjacent roads. Other examples of bad practice causing inconvenience to the public include:

- delivery vehicles queuing outside the site with their engines running before the site opens

- inconsiderate parking, blocking roads
- large vehicles reversing, causing delays to other road users
- vehicles being offloaded on the public highway/footpath, blocking traffic and inconveniencing pedestrians.

According to Allen *et al.* (2004), 26.25% of the 23,074 goods transported by 'British HGVs' operating 'within' or 'entering' Greater London consisted of 'building materials'. Interviews conducted with client representatives at Heathrow Airport during the DTI research project revealed that construction vehicle movements were reduced by up to 50% after the HCC was established, suggesting that the number of goods lifted by HGVs operating in Greater London could be reduced to 3029.

The single point of contact provided by the operators of a CC also increases the control of the Local Authority and the site. For instance, the CCM enhances the Local Authority's ability to influence the type of vehicle used (e.g. lower emission LPG), the delivery route and its timing.

Another major advantage of a CC is that security is improved because deliveries are made by a regular team of people, driving easily recognisable vehicles displaying the CC's logo. The traditional method, whereby numerous unmarked vans are seen arriving at and departing from the site may conceal thieves or criminals intent on targeting site neighbours.

Environment

The CCM reduces CO_2 emissions by reducing the number of deliveries normally required when suppliers deliver materials directly to a site. Figure 11.22 compares CO_2 emissions between the traditional route of materials delivery with the 'construction consolidation' method for the period of November 2001 to July 2003. For example:

Number of deliveries (from the HCC) × distance (to/from site 16 kms) × 0.502 kg CO_2

The number of deliveries is recorded at the end of each shift. A distance of 16 km (round trip) has been set from the HCC to the central terminal area and the amount of CO_2 emitted by a 7.5 ton vehicle is 0.502 kg per kilometre when operating at low speed. Although the amount of CO_2 emitted is inaccurate because figures are based on a 7.5 tonne vehicle and in reality a range of sized vehicles are used, the same is true for vehicles delivering materials to the HCC, therefore these are only indicative of the savings being made.

There is a direct correlation between the number of deliveries received and the potential to reduce CO_2 emissions. Savings are realised by the increased use of delivery vehicles on the final leg of the materials journey

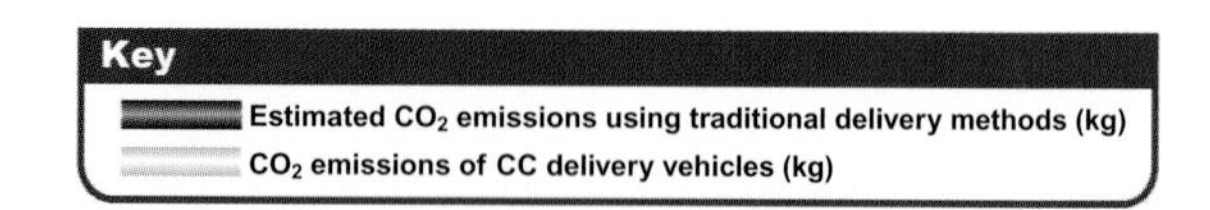

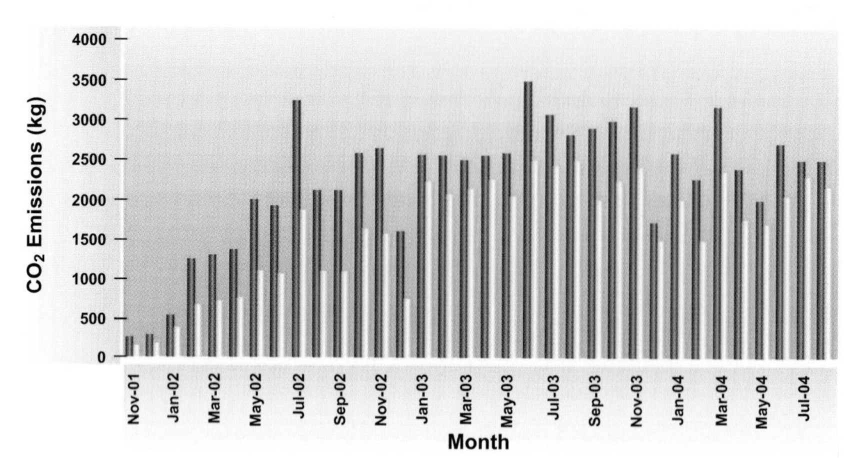

Figure 11.22 Comparison of carbon dioxide emissions between using traditional delivery methods and the consolidation centre delivery service

to site and are dependent on the deliveries' potential to be 'consolidated'. For instance, if ten deliveries arrived at the CC and each was sent out individually (on ten separate journeys to site), no saving would be made. This is evident when comparing July 2002 with April 2003. Even though approximately the same numbers of deliveries were received in both, the deliveries received in April offered less potential for consolidation, therefore the environmental benefit was less.

Conversely, it is apparent from comparing December 2002 with April 2003 that other factors influence the potential for consolidation and the extent of environmental benefits. For instance, in December 2002, CO_2 emissions were reduced by 42.4% compared with April 2003, when CO_2 emissions were reduced by just 11.8% as a result of an improved consolidation of loads.

Other factors which affect the HCC's ability to consolidate numerous consignments on to one delivery to site include the:

- throughput of materials and equipment
- size of the CC's vehicles compared with the materials supplier's vehicles
- size of individual consignments
- trade contractor's ability to plan work packs in advance
- effectiveness of the communication between the trade contractor and the CC
- level of urgency placed on the material or equipment being delivered.

Average monthly CO_2 emissions were reduced by 750 kg and the total reduction over the period represented in Figure 11.22 equates to 15 tonnes of CO_2. It was apparent that larger quantities of materials were being delivered to the HCC than would have been possible if deliveries were destined for site because of space constraints. Therefore, further consideration and recognition should be made to the effect that the HCC has on the consolidation of deliveries upstream in the supply chain.

The principal driver for BAA developing the HCC was to increase control over the materials delivery process. It was appreciated that a single logistics service provider increases the level of control and accountability. Airside construction vehicle movements were reduced by up to 50% at Heathrow Airport. A CC's ability to reduce vehicle movements corresponds with its ability to consolidate numerous consignments onto one vehicle and is intrinsically related to the increased use of delivery vehicles. Reducing the number of vehicles delivering to site also reduces security issues.

The social benefits of good supply chain management and logistics techniques

One of the recent business drivers for implementing responsible SCM and logistical best practice has been influenced by the emergence of corporate social responsibility (CSR), or corporate responsibility (CR) as it is also referred to. CSR is a fast-developing phenomenon with industry in general and with the construction industry in particular because of the significant impact that construction projects have on their diverse stakeholders. The Business for Social Responsibility's definition of CSR is 'achieving commercial success in ways that honour ethical values and respect people, communities, and the natural environment'. Bloom and Gundlach (2001) define CSR as, 'the obligations of the firm to its stakeholders – people and groups who can affect or who are affected by corporate policies and practices. These obligations go beyond legal requirements and the company's duties to its shareholders.'

Construction projects are often located amidst residential and commercial neighbours and thus create additional logistical constraints. CSR relates to the ethical conduct of business and is about exceeding legal minimum requirements.

The recognised pillars of CSR that are promoted by Business in the Community (www.bitc.org.uk) are:

- Workplace: issues concerning employer and employee relationships, health, safety and welfare, career development and equal opportunities, diversity and employment conditions.
- Marketplace: issues concerning supply chain partner relationships and ethical business practice.

- Environment: having regard for the environment by implementing sustainable business policies, practices and procedures, and minimising the negative impact of business operations, especially the corporate carbon footprint.
- Community: providing financial, in-kind and employee volunteering support for communities and charities.

Barthorpe and Gleeson (2004) contend that the fragmented and diverse nature of the construction industry illustrates why there is an inconsistent level of CSR demonstrated by developers and contractors in the UK. Although there are some exemplary developers and contractors, the overall impression is that CSR is performed on a rather ad hoc basis. An excellent CSR framework, specific to the UK construction and property industry, however, exists in the form of the Considerate Constructors Scheme's (2006) *Code of Considerate Practice*. Contractors and developers apply this code of practice on an entirely voluntary basis, setting standards of behaviour well above the minimum statutory levels of compliance. In this respect, parallels can be drawn with the tenets and philosophy of CSR. Implementing and integrating CSR into mainstream business practice is not a philanthropic or altruistic gesture. There are clear business benefits to be gained by implementing CSR, including:

- increased intangible assets and brand value
- increased investor confidence and access to finance and markets
- increased client base and business opportunities
- improved relationships with and respect from clients and other stakeholders
- increased recognition as a responsible 'corporate citizen'
- improved 'triple bottom line' (economy, society and environment)
- improved recruitment and retention of employees
- improved relationships with the community – 'licence to operate'
- increased competitive advantage.

Socially responsible supply chain management and logistics

SCM and logistics affect different sets of stakeholders. The former has a greater impact upon society at large, whereas the latter affects society at a local or regional level. For example, through SCM it is possible to appoint suppliers based on the environmental merits of their production techniques, certifying that the materials they use originate from sustainable sources and that they conduct their business in an ethical manner. Whereas logistics has a greater effect on the community adjacent to the site, which might be disrupted as a result of noise pollution, congestion and inconsiderate parking or unloading techniques. This difference is

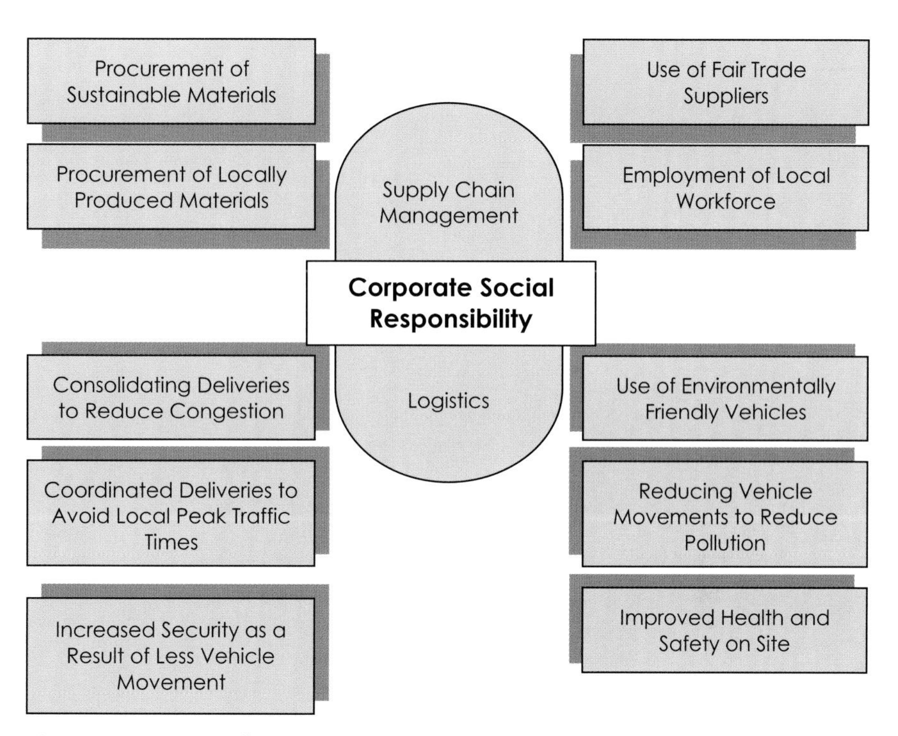

Figure 11.23 Socially responsible supply chain management and logistics techniques

reflected in Figure 11.23, which considers socially responsible SCM and logistics techniques.

Although SCM and logistics can be considered entirely separate disciplines, they share many features. For instance with supply chain effect on logistics it is possible to reduce vehicle movements and cost by eliminating non-value-adding members of the supply chain or by appointing suppliers based on their proximity to site, therefore reducing the environmental impact associated with materials delivery. Similarly, the logistics effect on the supply chain is influenced by a logistics strategy that promotes the use of reusable, purpose-built containers to minimise packaging waste.

Considering the future application of the consolidation centre methodology

Two flow charts have been designed to consider the future application of the CCM. The first flow chart (Figure 11.24) considers project-specific constraints, such as limited storage space, to decide whether a CC is required. The second flow chart (Figure 11.25) considers factors, such as the location of the site, to prescribe the most suitable CC model. Using

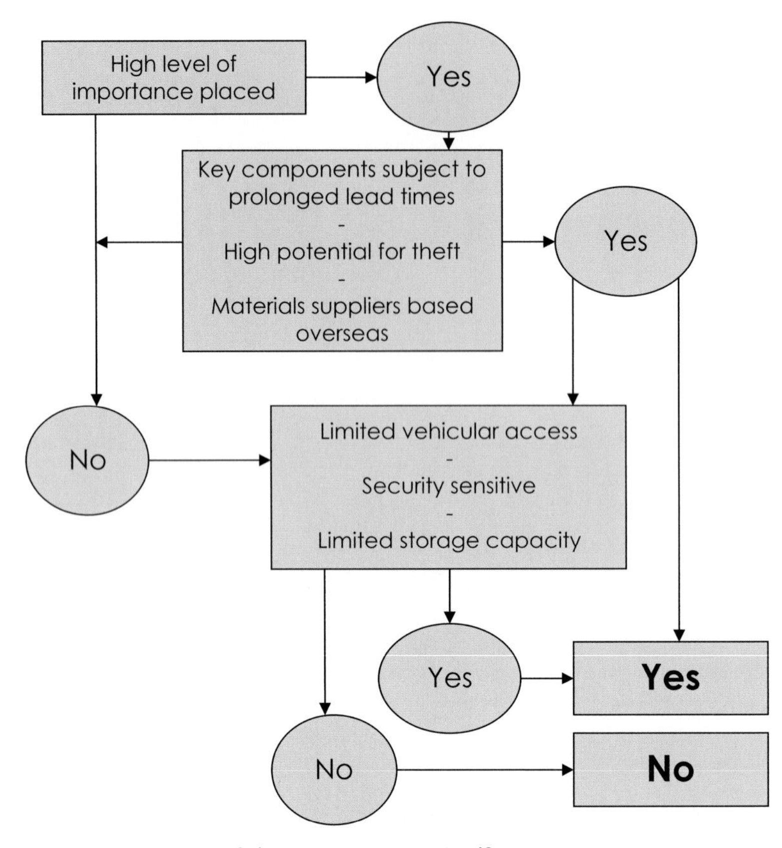

Figure 11.24 Is a consolidation centre required?

this flow chart assists the decision-making process for clients or contractors who wish to adopt the CCM. It considers the logistical constraints which influence the vast majority of sites; however, a degree of flexibility should be applied when considering the optimum solution. This might involve using a hybrid CC with characteristics from two or more of the proposed CC options.

Option 1: The concealed consolidation centre

The concealed consolidation centre, so called because it is concealed within the boundary of the site hoardings, is the most basic form of a CC and is operated by the site's principal contractor. Typically, it has the following characteristics:

- a site-based temporary warehouse unit approximately 500 m^2
- uses ten principal contractor logistics employees

- a requirement to receive and distribute materials and equipment only during the day
- operating hours determined by the Local Authority
- a basic paper-based system used to control inventory
- telescopic forklift truck, hand pallet transporters and trolleys
- funding from the client; however, some costs can be reclaimed by omitting logistics costs from tenders.

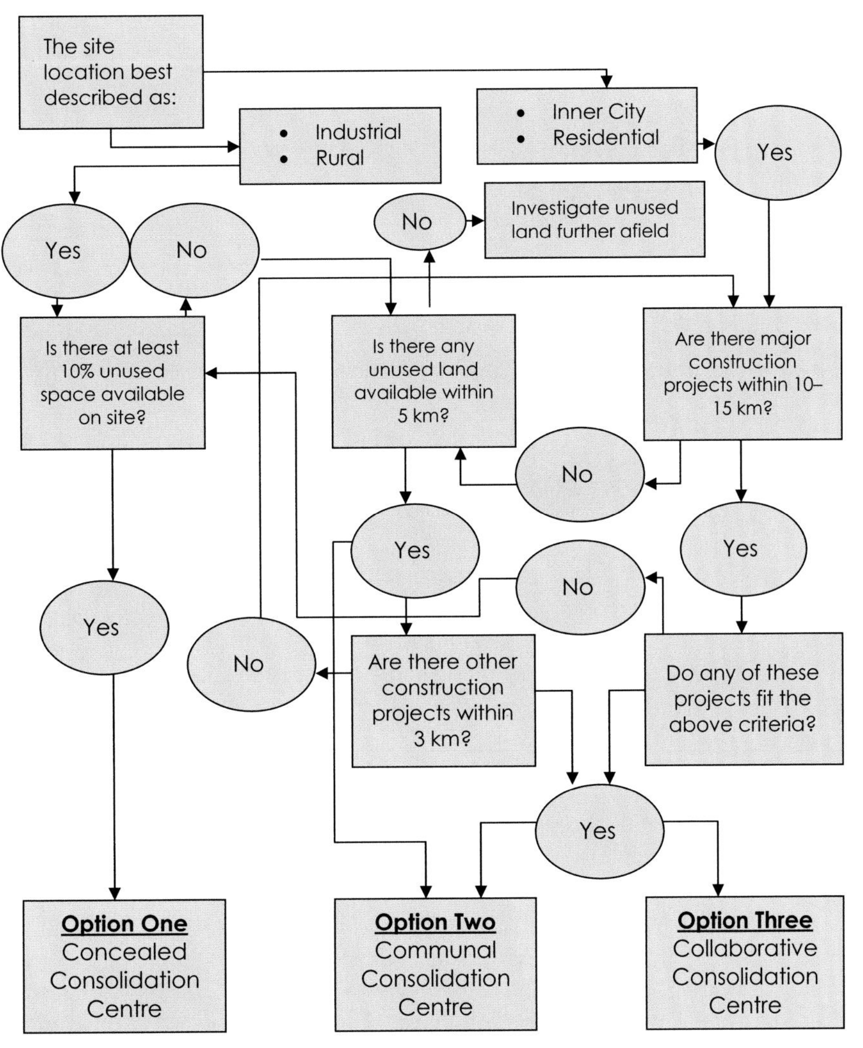

Figure 11.25 Flow chart to select the most appropriate consolidation centre option

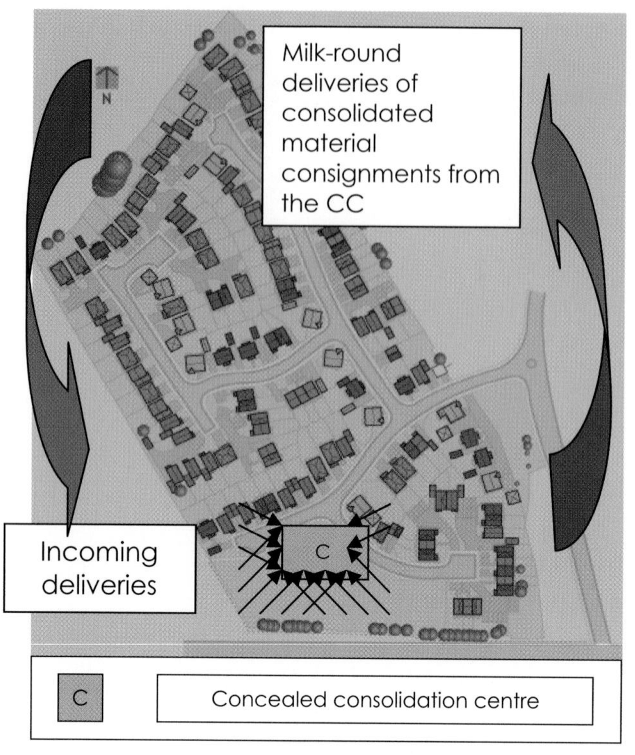

Figure 11.26 The way in which materials suppliers deliver materials and equipment directly to the consolidation centre

The concealed consolidation centre's main purpose is to:

- provide a production buffer for suppliers
- provide appropriate levels of material stock for contractors
- provide a secure warehouse facility
- reduce damage to equipment and materials
- improve health, safety and welfare standards
- reduce waste and facilitate the reuse/recycling of packaging
- maximise productivity on site by using specialist logistics operators.

Figure 11.26 illustrates the way in which materials suppliers deliver materials and equipment directly to the CC. Items are stored for up to one week and delivered to the workface on a JIT basis.

As materials suppliers are required to deliver to the CC that is based on site, the concealed consolidation centre does not offer any environmental benefits in terms of reduced congestion or CO_2 emissions. However, its purpose is to maximise the productivity of the skilled workforce, which it achieves by eradicating the need for tradespeople to leave their work to offload equipment and materials. Logistics operatives employed by the CC are responsible for receiving and distributing all consignments. This includes unloading, checking and storing the equipment and materials received, then selecting and transporting the equipment and materials from the store to the required destination as and when required.

Option 2: The communal consolidation centre

The communal consolidation centre is so named because it serves numerous single-client/contractor projects at the same time. Typically, it has the following characteristics:

- located reasonably close to the construction projects (maximum 5 km away)
- a factory-style building approximately 1000 m^2
- operated by the principal contractor or a specialist logistics contractor
- using up to 20 employees
- operating flexible hours, capable of receiving deliveries during the day and distributing materials at night
- capable of serving a number of sites within a relatively small area
- using a simple Excel-based spreadsheet system to control inventory
- resources include telescopic forklift truck, reach and tier forklift truck, a number of vehicles up to 27 tonne rigid-bed lorry, hand pallet transporters and trolleys
- principally funded by the client; however, some costs might be reclaimed by omitting logistics costs from tenders.

The communal consolidation centre's main purpose is to:

- provide a production buffer for suppliers
- provide appropriate levels of material stock for contractors
- provide a secure warehouse facility
- reduce damage to equipment and materials
- improve health, safety and welfare standards
- reduce waste and facilitate the reuse/recycling of packaging
- maximise productivity on site by using specialist logistics operators
- reduce materials congestion.

Figure 11.27 illustrates an example of a communal consolidation centre that has been set up to serve a number of single-client/contractor projects in an area. It also illustrates the method used by materials suppliers. Upon arrival, the materials would be checked by logistics operatives. As this type of CC would serve a number of sites, it may be necessary to provide each site with its own storage area; however, generic materials such as plasterboard, bricks, blocks, sand and cement could be stored more centrally. Alternatively, if space did not permit this form of storage, it would be possible to operate a communal consolidation centre where different sites materials and equipment are integrated. This would require more robust labelling methods and the use of a WMS.

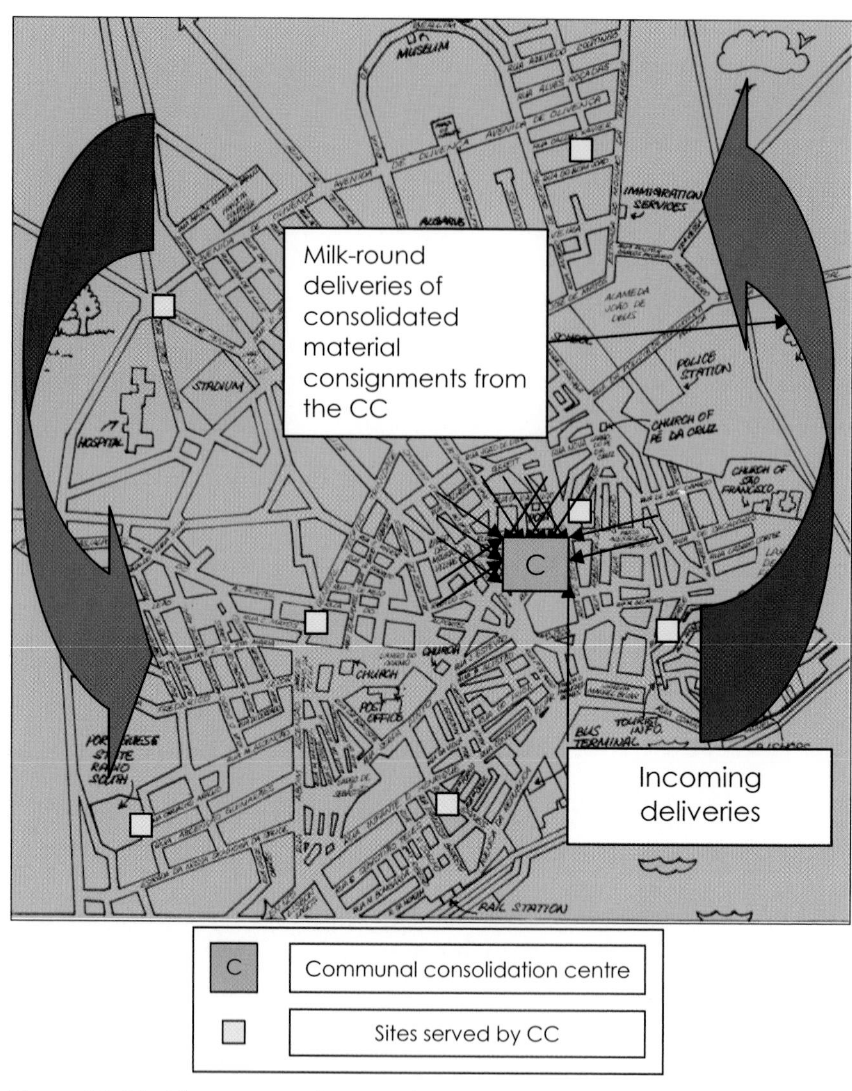

Figure 11.27 The communal consolidation centre materials flow process

Option 3: The collaborative consolidation centre

The collaborative consolidation centre receives its name from its demand for collaboration between different clients and contractors. It is the largest of the three options and its use might even be specified as a planning pre-condition by the Local Authority. Given the extent of the non-tangible benefits listed below under 'main purpose', this form of CC is the most likely to receive government funding.

A collaborative consolidation centre is characterised by:

- being located up to 25 km away from the sites
- being a purpose-built, permanent construction logistics centre of approximately 4000 m^2
- being operated by a specialist logistics contractor
- using around 50 employees
- being operational 24 hours a day, seven days a week and being capable of receiving deliveries during the day and distributing materials at the night
- being capable of serving all major construction projects for a town or city
- using a WMS to control inventory
- using a range of mechanical resources, including telescopic forklift trucks, a number of vehicles up to articulated lorry size, hand pallet transporters and trolleys
- being financed by a handling charge, e.g. on a cost per pallet basis.

The main purpose of a collaborative consolidation centre is to:

- provide a production buffer for suppliers
- provide appropriate levels of material stock for contractors
- provide a secure warehouse facility
- reduce damage to equipment and materials
- improve health, safety and welfare standards
- reduce waste and facilitating re-use/recycling of packaging
- maximise productivity on site by utilising specialist logistics operators
- reduce materials congestion
- be a centre of excellence for logistics training and development
- increase accountability and enhance the image of the construction industry.

Figure 11.28 provides an example of a collaborative consolidation centre that has been set up to serve several construction sites located in London. Its proximity to the strategic transport network is fundamental for access and egress, both for the CC and the materials suppliers' vehicles.

Depending on the size of the area served by the CC, the number and nature of construction sites and the geographical complexity of the area, it may be necessary to operate a number of this type of CC. For example, regional CCs might be used to serve the northern, southern, eastern and western regions of a city.

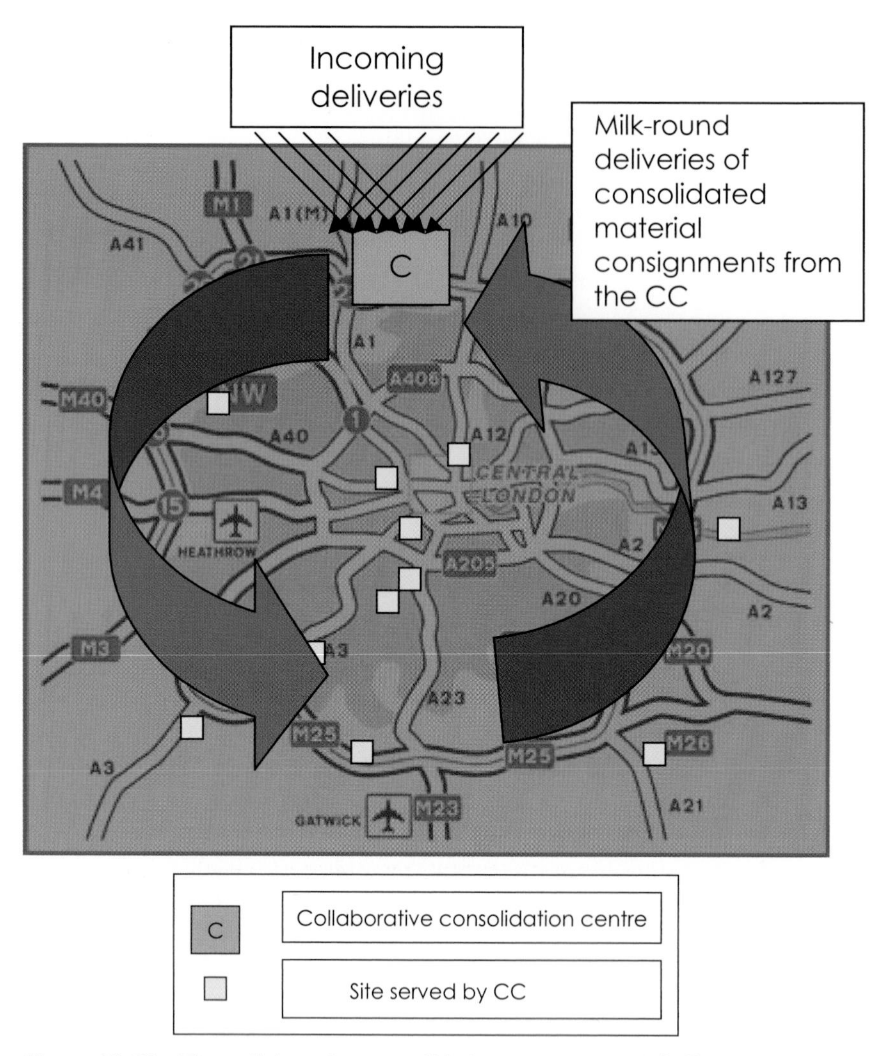

Figure 11.28 The collaborative consolidation centre materials flow process

Important considerations for the future

Each version of the CC would itself be subject to planning conditions and as such may have restrictions imposed which curtail certain activities such as night-time deliveries as demonstrated by the Colnbrook Logistics Centre, which serves Heathrow Terminal 5. Planning restrictions are most likely to affect the communal and collaborative versions of the CC because of the increased congestion in the vicinity and the associated demand on the adjacent infrastructure.

The HCC has demonstrated the effectiveness of its palletisation strategy, without which manual handling and the risk of associated injuries are increased. The palletisation strategy facilitates the optimum use of storage capacity within the CC and of the vehicles whilst in transit. More importantly in terms of the future of the CCM, it provides a useful link with materials suppliers that should be used to influence the way in which materials are delivered to site. In each of the three models of the CC proposed, materials suppliers could further enhance the benefits by issuing work packs designated for a particular area of a building which coincides with a working drawing. This highlights the extent to which the CCM could revolutionise the way in which buildings are constructed and demonstrates the need to implement integrated design considerations and apply a holistic change in the construction supply chain process.

To further enhance CC vehicle utilisation rates, operators should consider providing their sites with a waste collection service. This would prevent their vehicles returning empty, further reducing congestion by providing a central collection point for refuse contractors. Furthermore, this service could be extended to salvage surplus materials, providing an opportunity to reuse previously wasted material, returned from site. Such materials might be offered for sale to the public or donated to Local Authorities or charities for civic or community-enhancing projects.

The argument for using CCs is compelling and would become more routinely adopted if clients and developers of inner-city projects in particular paid more attention to the associated logistical constraints. Local planning authorities have a significant role to play too and should consider granting planning permissions conditional on being satisfied that the project planning application submissions include a full logistics strategy and commitment to implement a CC where appropriate.

References

Allen J, Browne M and Christodoulou G (2004) *Freight Transport in London: A summary of current data and sources*. Transport for London, London.

Barthorpe S and Gleeson J (2004) 'Implementing corporate social responsibility through the framework of the Considerate Constructors Scheme'. Proceedings of the Corporate Social Responsibility and Environmental Management Conference, University of Nottingham, June 2004.

Bocij P, Chaffey D, Greasley A and Hickie S (2003) *Business Information Systems: Technology, development and management for the e-business*. Pearson, Harlow.

Bloom PN and Gundlach GT (2001) *Handbook of Marketing and Society*. Sage Publications, Thousand Oaks, CA.

Collins J (2004) Case builds for RFID in construction, http://www.identecsolutions.com/pdf/IDENTECSOLUTIONS_RFIDConstruction_RFID%20Journal.pdf, [accessed 20th March 2006].

Considerate Constructors Scheme (2006) Home page, http://www.ccscheme.org.uk, [accessed 20th March 2006].

Department of Trade and Industry (2004a) *Construction Logistics Consolidation Centres: An examination of new supply chain techniques: Managing & handling construction materials*. DTI, London.

Department of Trade and Industry (2004b) *Consolidating Construction Materials Transferring the Methodology: Final Report*. DTI, London.
HMSO (1998) Working Time Regulations 1998, SI 1998/1833, http://www.opsi.gov.uk/si/si1998/19981833.htm#10, [accessed 22nd January 2010].
Johnson G and Scholes K (1999) *Explaining Corporate Strategy: Text and cases*, 5th edn. Pearson Education, Harlow.
Rother M and Shook J (2003) *Learning to See: Value stream mapping to create value and eliminate muda*. Lean Enterprise Institution, Brookline, MA.

Chapter 12
Case Studies

Introduction

The origins, theory, development and practitioner guidance on logistics has been provided in the previous chapters of this book. This chapter provides seven industry case studies, demonstrating different aspects of logistics in practice that have been used successfully on major construction projects in the UK, including some award-winning industry examples.

Case Study 1: Construction Logistics: The Heartbeat of a Project

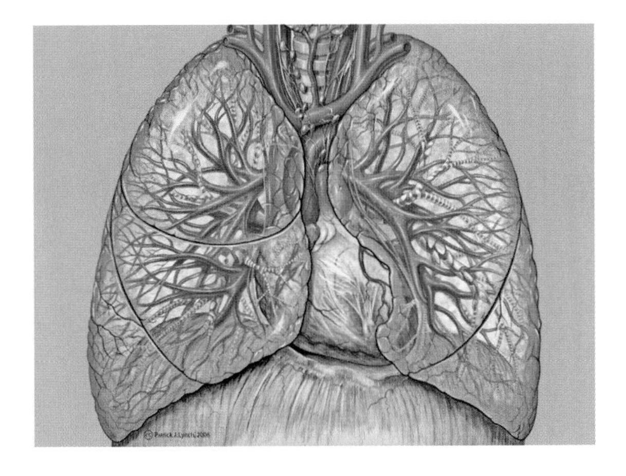

In this first case study, the anatomy of the human body has been used as a metaphor to demonstrate the vital contribution that logistics makes to the successful management of a construction project. Different logistics applications from several construction projects have been demonstrated with actual examples of vehicle movement and carbon dioxide emissions call-off data sheets and histograms provided.

By applying a little bit of imagination, it is relatively easy to use the human anatomy as a metaphor and compare the healthy functioning of the lungs, blood vessels, heart, muscles etc. with the efficient flow of materials, removal of waste, expiration of CO_2 and productivity of a construction project.

Case Study 1: Construction Logistics: The Heartbeat of a Project

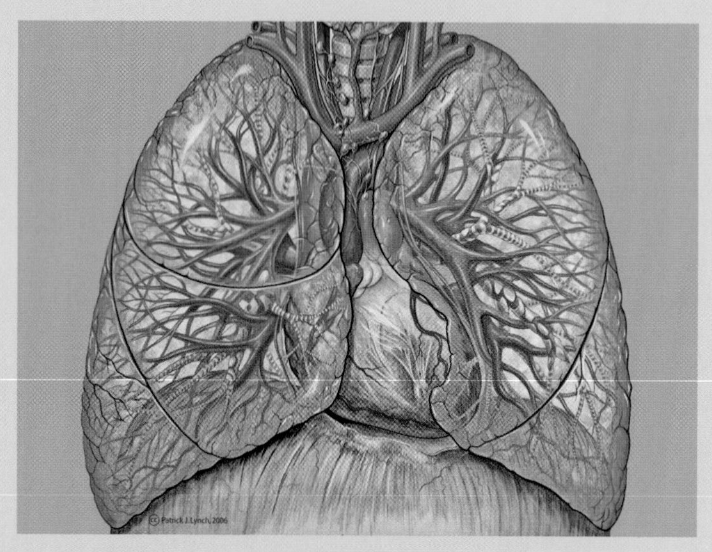

Human pulmonary system

What is construction logistics?

Imagine a human body. To work and grow efficiently, lungs are needed to capture and hold oxygen (and expel CO_2), a network of blood vessels is crucial to deliver oxygen, nutrients and hormones direct to muscles (and to remove the CO_2 waste product), and a heart is essential to regulate and manage the flow of blood around the body to maximum effect.

Now imagine a big and complex construction project. In order for it to grow to schedule and budget, its building materials (the lifeblood of the site) must be delivered on time and then transported to the workface where the skilled trades need them: the construction site needs a heartbeat too. Waste product has to be removed and expelled from the project to prevent accidents and inefficiencies. The whole process has to be regulated and managed to make the most of hoists, cranes and available labour.

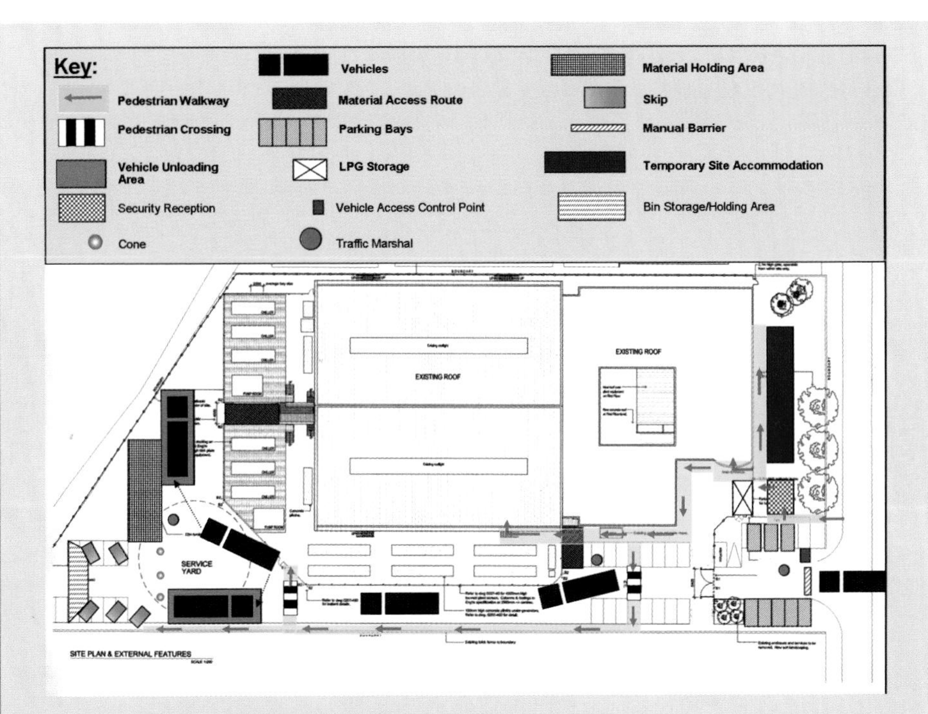

Logistics traffic/materials handling project plan

This is called 'construction logistics'.

Construction logistics is as vital to the successful construction of projects, such as Terminal 5, as a strong heartbeat is to an Olympic athlete.

And Wilson James is that strong heartbeat, keeping materials flowing.

Anatomy of a construction project: Terminal 5

£4.3 billion project • 5-year build plan • heavily constrained site • immovable deadline for completion • 2 main terminal buildings • 60 aircraft stands • air traffic control tower • 4,000-space multi-storey car park • 600-bedroom hotel • all served by a new spur road from the M25 • 3,000 construction personnel on site at peak • 200 HGV deliveries a day at peak.

Wilson James Ltd (WJ) was employed by BAA on Terminal 5 in 2003 as its construction logistics contractor, undertaking site set-up and identifying the logistics requirements to service the site, working within BAA's Site Establishment Team (SET).

The focal point for all construction material deliveries was the Colnbrook Logistics Centre (the project's 'lungs'). Each of the various projects and contractors had access to a software booking system (the 'heart'), managed by WJ, which collated the teams' demands and allocated booking times into the centre. A report of the next day's deliveries was issued each afternoon to security and the booking-in clerk. Materials (the 'blood vessels') were then delivered either just before or on the day they were required. Consolidated loads were called off by the project using the same software for cross-referencing and the load was delivered to the workface to a pre-arranged time.

Case Study 1: Construction Logistics: The Heartbeat of a Project (*Continued*)

Heathrow airport's Terminal 5 under construction

This process eliminated the need for extensive lay down areas for materials. Implementing a robust logistics strategy at Terminal 5 realised significant savings. The increased reliability and efficiency in the use of materials led to an increase in productivity levels from a traditional build of 55–60% to 80–85% with associated cost savings in the region of 2.5%.

London Construction Consolidation Centre: A capital idea

95% delivery performance • site labour productivity up 30 minutes per person, per day • 15% reduction of materials wasted • £200,000 of reusable materials on one project recovered • 2-hour reduction in supplier journey times • 68% reduction of construction vehicles entering the City of London • 75% reduction of CO_2 emissions • 25% reduction in accidents.

The LCCC situated in Silvertown, London

In 2005, the London Construction Consolidation Centre (LCCC) was born out of the successful implementation of the consolidation process at Heathrow Airport. WJ partnered with Transport for London, Stanhope and Bovis Lend Lease to pilot the centre in SE16, serving four key City of London projects. At the end of the pilot in 2007, the LCCC was shown to have made a considerable impact on all four projects: increasing productivity and reducing waste, cost, unnecessary journeys for delivery vehicles and, of course, their CO_2 emissions.

Materials are stored safely on racking at the LCCC

Now with the pilot programme complete, WJ continued to own and operate the only independent construction consolidation centre in the UK. The LCCC moved to bigger premises in Silvertown, with even better access to the throbbing heart of London, and immediately started servicing the material handling and distribution needs of Skanska (Barts Hospital), Structuretone (155 Bishopsgate), Overbury (RBS Bankside) and Bovis Lend Lease (Central St Giles). Trade contractors themselves woke up to the benefits of using the LCCC to regulate the flow of their materials to projects across London: Mag Hansen, Hall & Kay, Levolux and Mivan, to name just a few.

Case Study 1: Construction Logistics: The Heartbeat of a Project (*Continued*)

The graphs show how over the course of six months only (March–September 2008) the LCCC removed unnecessary vehicle trips to site, regulating the flow of materials and reducing the carbon footprints for the Skanska Barts Hospital Project.

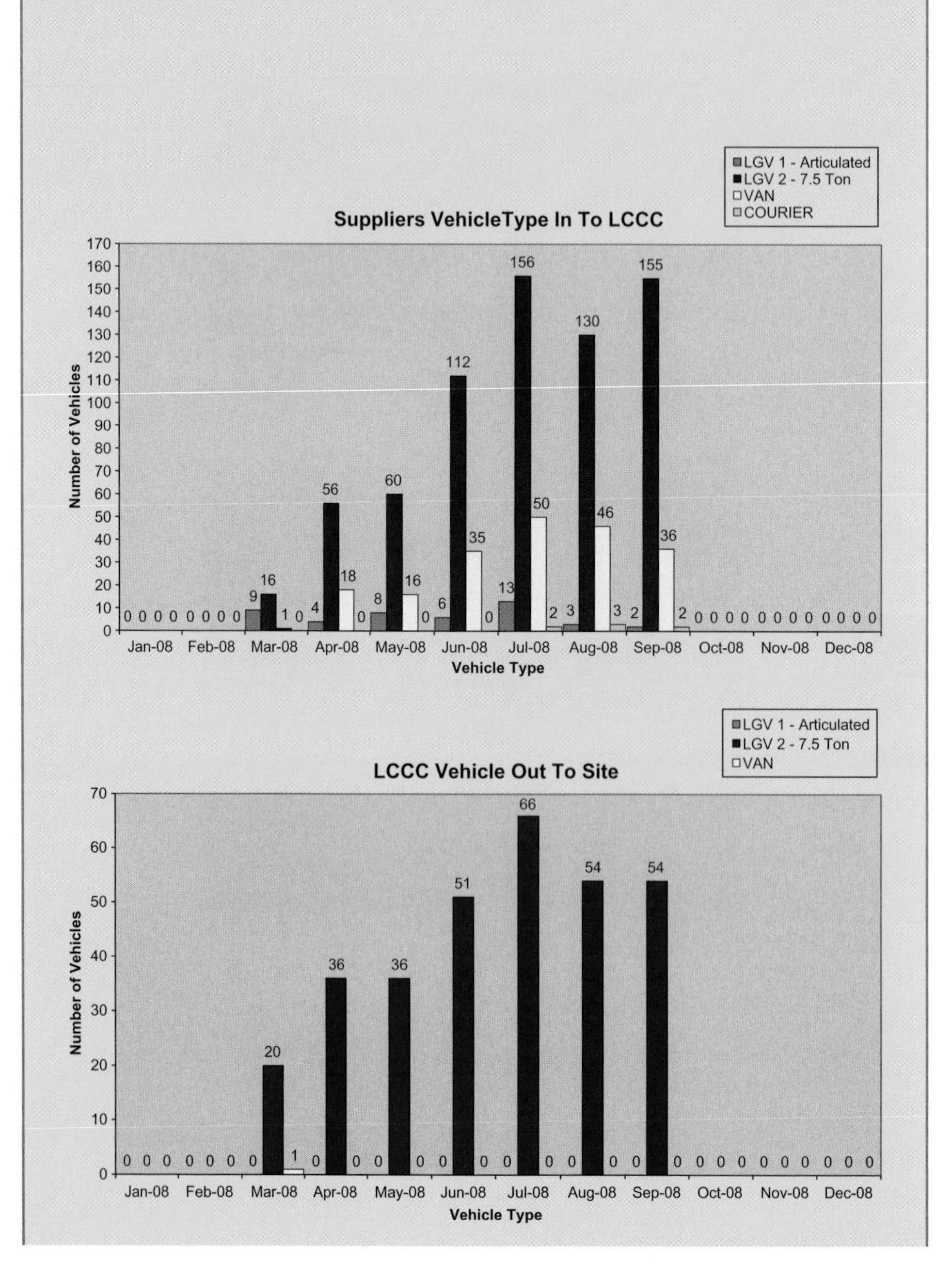

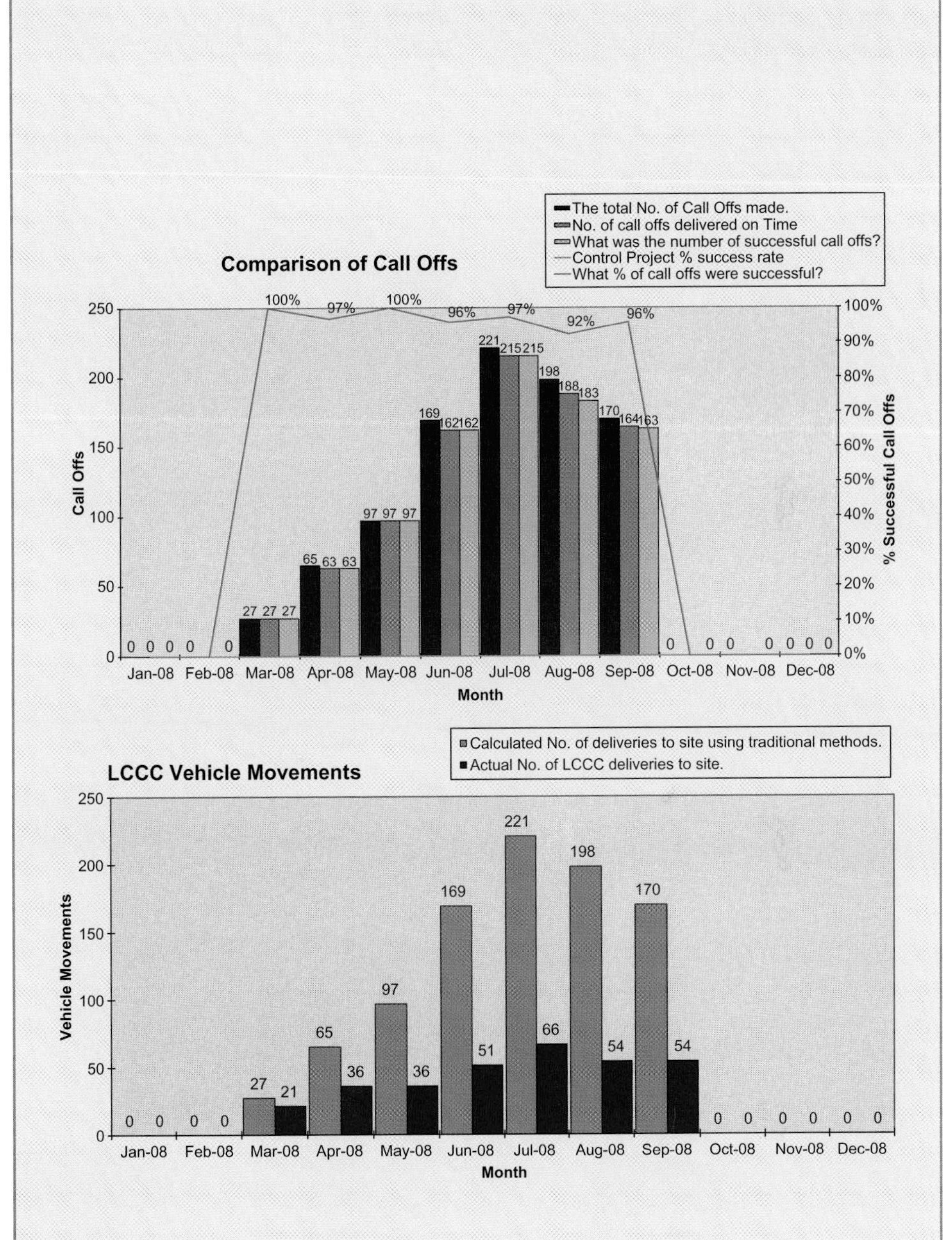

Comparison of Call Offs
The total No. of Call Offs made.
No. of call offs delivered on Time
What was the number of successful call offs?
Control Project % success rate
What % of call offs were successful?
Call Offs
% Successful Call Offs
Month
Jan-08 Feb-08 Mar-08 Apr-08 May-08 Jun-08 Jul-08 Aug-08 Sep-08 Oct-08 Nov-08 Dec-08
100% 97% 100% 96% 97% 92% 96%
LCCC Vehicle Movements
Calculated No. of deliveries to site using traditional methods.
Actual No. of LCCC deliveries to site.
Vehicle Movements
Month

Case Study 1: Construction Logistics: The Heartbeat of a Project (*Continued*)

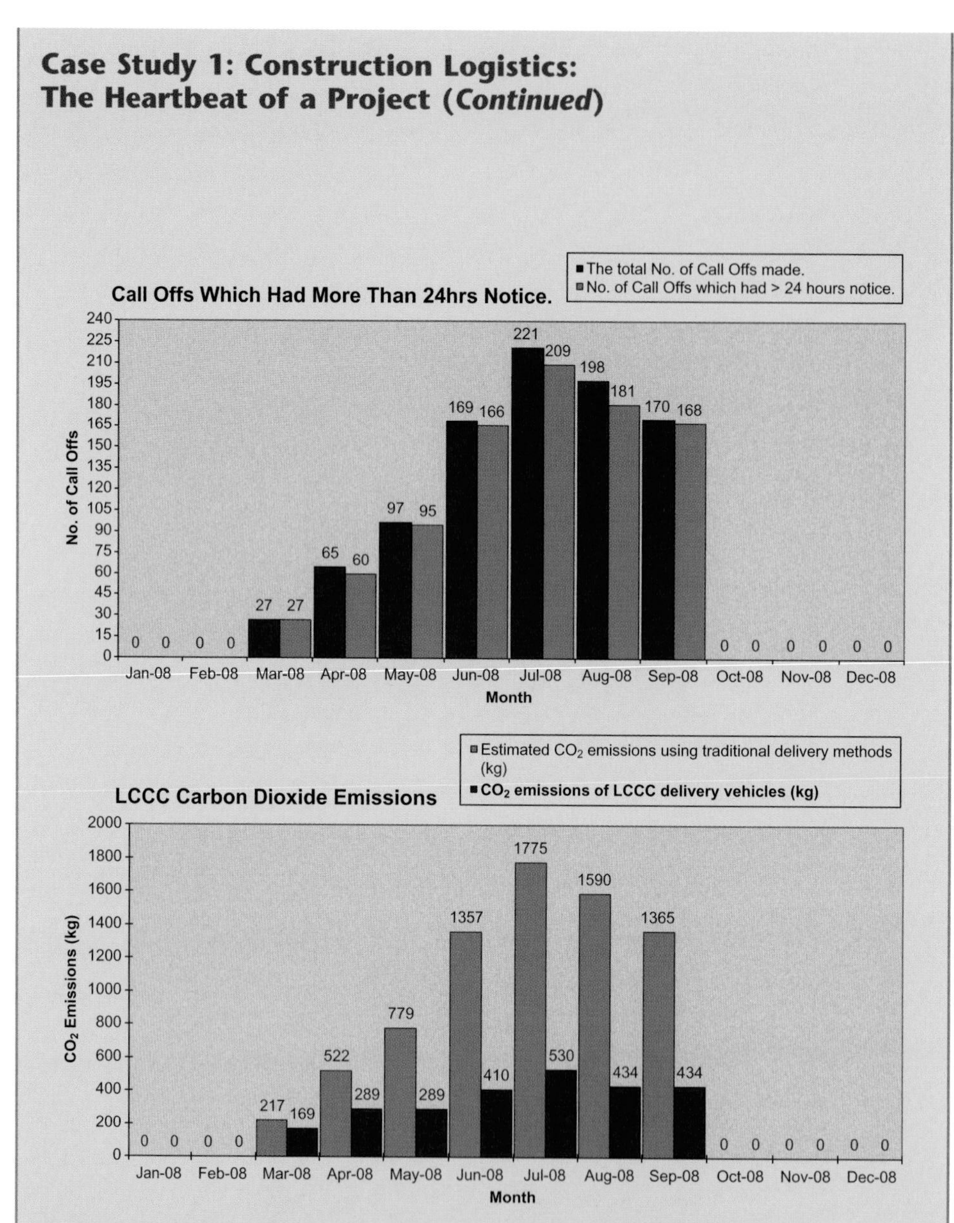

Skanska Barts/LCCC vehicle and call-off data, March–September 2008.

Section 3

The prognosis

In June 2008, BAA awarded WJ the Construction Logistics Integrator role, tasked with delivering across BAA's entire £4.3 billion Capital Project build programme until 2013 the following services:

- materials management
- waste management
- security services
- construction vehicle routing
- traffic management
- facilities management
- canteens
- car parking and bussing
- workers' accommodation helpline.

Effective logistics planning requires having access to detailed plans

Indeed, WJ is increasingly working alongside clients such as BAA, Stanhope, AWE, AstraZeneca and SnOasis to incorporate construction logistics into their plans at an early stage, to ensure their projects grow efficiently and healthily to maturity.

But the effective application of construction logistics can also help those projects that, for whatever reason, are battling the odds to win through. WJ was employed by Westfield to manage the flow of materials and waste for the fit out of Europe's biggest shopping centre, Westfield London. Time was short and in just 16 weeks WJ was able to:

- offload 10,931 delivery vehicles
- unload 619 oversized deliveries
- deliver 49,046 pallets to shop units.

During the last week alone, WJ managed 926k sq/ft of active fit-out, dealing with 288 tenants, over 367 units. On 30th October 2008, Westfield London opened its doors on time to great acclaim.

Case Study 1: Construction Logistics: The Heartbeat of a Project (*Continued*)

Artist's impression of the Westfield London shopping centre

Now that the economy is unwell and construction is feeling queasy, a liberal dose of construction logistics could be just what the doctor ordered.

Case Study 2: Prescription to Reduce Waste: AstraZeneca

An insight into waste management is provided in Chapter 3. The introduction of the Site Waste Management Plans Regulation 2008 made it a requirement that all construction projects over £300,000 value must have a site waste management plan (SWMP). The AstraZeneca project complied with this legislation a year before its statutory implementation. This £63.5 million process research and development laboratory in Macclesfield, UK faced many challenges, not least the difficulties of controlling the flow of materials and the removal of waste due to the limited amount of storage available on site.

Only 8–10 dedicated logistics operatives were employed on this project, yet significant waste management achievements and impressive targets were attained. A reward and penalty scheme was in place for the site personnel to optimise the degree of waste segregation, reuse and recycling of construction materials. This project won the Recycling Target Success Category at the National Recycling Awards in 2008.

Case Study 2: Prescription to Reduce Waste: AstraZeneca

Recycling and Environmental Manager, Phil Harrington receives a Lifetime Achievement Award at the 2009 National Recycling Awards

AstraZeneca's £63.5 million process research and development (PR&D) laboratory in Macclesfield has acted as a test bed for a new approach to construction logistics for the pharmaceutical giant.

As with most construction projects, one of the biggest, but often least controlled, issues is the storage and flow of materials to and waste product from the workface. In addition, AstraZeneca (AZ) had both a very limited amount of storage on site and a high quota of prefabricated elements to the build.

It was keen to comply with the new site waste management plan (SWMP) legislation a full year ahead of its statutory implementation.

Wilson James was able to help AZ achieve its goals by establishing four separate compounds on site for the management of materials and by preparing and implementing a full SWMP.

The overall target was to achieve maximum sustainable recycling, recovery and reuse of construction materials. This was achieved by:

- thorough pre-contract planning involving all contractors
- following SWMP regulatory principles
- segregating at the workface waste streams in colour-coded containers
- devising and implementing a comprehensive waste management and materials handling training programme for all site personnel, with reward and penalty schemes in place
- dedicated bulking up areas to maximise loads
- direct journeys to real processors of recyclables rather than loading it into a mixed-use skip
- accurate/timely measurement of all recyclables to produce reliable information and key performance indicators
- the use of competent waste personnel for the whole process, driving reuse as a principal target, introduction of a colour-coded scheme to segregate various waste materials into their different streams, and prevent cross-contamination.

Operational achievements

- Over 280 fume cupboards fully assembled off-site, delivered to the Macclesfield storage facility area with no damage.
- Over 500 prefabricated services modules, up to 6m long, were delivered just in time with no damage or adverse effect on site activities.
- Correct materials handling equipment sourced, thus minimising costs.
- Little or no site congestion, due to managed delivery schedules, resulting in no traffic-related disruption to AZ infrastructure or personnel.
- 99.24% of the overall waste produced was either recycling, recovered or reused.
- 70% of waste was recycled, delivering cost efficiencies by optimising transport providers for each waste stream.
- A high standard of housekeeping was maintained throughout the fit-out phase.
- Only 8–10 logistics operatives were used, replacing 30 individuals cumulatively provided by trades.
- Site management personnel were freed up from sorting out materials deliveries, enabling them to focus on core activities.

Case Study 2: Prescription to Reduce Waste: AstraZeneca (*Continued*)

An example of the colour-coded scheme introduced to segregate different waste streams.

National Recycling Awards 2008

In November 2008, in recognition of the successful project waste management operation, AZ and Wilson James Ltd were announced winners of the Recycling Target Success Category at the National Recycling Awards.

Wilson James on-site team receives its trophy

Case Study 3: Consolidated Loads: The Answer to Space- and Access-constrained Sites

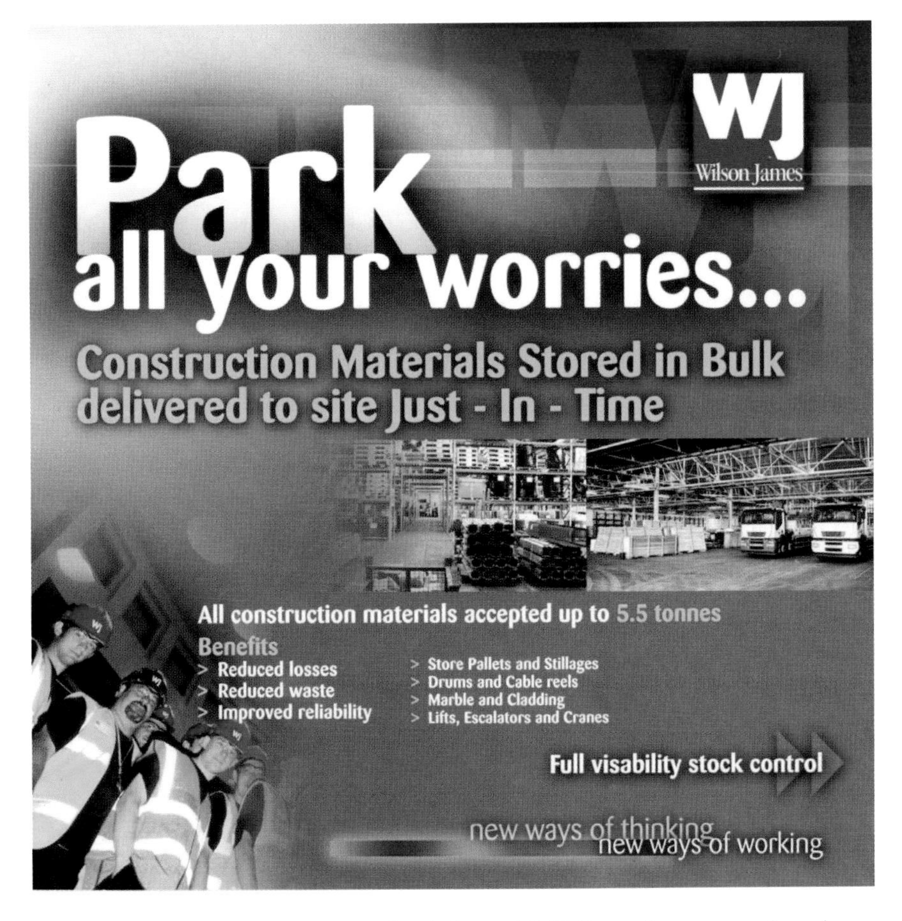

The consolidation centre methodology (CCM) and concept has been described in detail in Chapter 11. The London Construction Consolidation Centre (LCCC) featured in this case study is modelled on BAA's consolidation centre at London's Heathrow Airport. The LCCC is a modern facility incorporating properly trained people and efficient materials handling equipment to ensure the correct handling, temporary storage and timely delivery of construction materials and equipment from numerous suppliers to their point of use on sites.

The LCCC serves numerous construction projects in the London region simultaneously and is instrumental in significantly improving site productivity as well as providing environmental and social benefits by consolidating loads, which reduces the frequency of materials deliveries. Waste is also minimised and unnecessary packaging removed and recycled rather than sent to site.

Wilson James manages the LCCC and has won numerous supply chain, sustainability and innovation awards.

Case Study 3: Consolidated Loads: The Answer to Space- and Access-constrained Sites

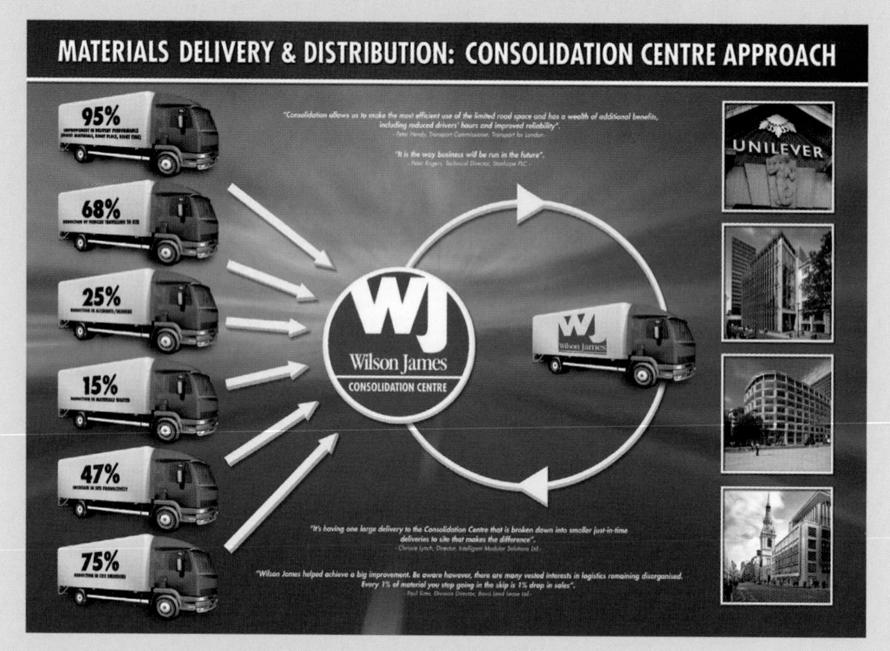

Construction consolidation centre model approach

The consolidation centre methodology (CCM) approach demonstrates the key role supply logistics has to play in preventing construction waste by delivering materials to site when needed. By serving a large project or a number of different jobs, the centres can reuse surplus materials, rather than send them to landfill.

The London Construction Consolidation Centre (LCCC), modelled on the BAA centre at Heathrow, is widely accepted as one of the most recent demonstrations of how lessons learnt from other industries can improve the performance of the construction industry.

The main purpose of the LCCC is to promote the efficient flow of construction materials through the supply chain to the actual points of use on projects. It is not a warehouse.

The centre aims to enhance construction sites' performance and reduce the impact on the environment, by reducing congestion, pollution and noise.

Key facts

- Reduction of materials waste of up to 15%.
- Achievement of delivery performance of 95% of goods delivered, right first time.
- Increased productivity of the site labour force of up to 30 minutes per day; which equates to 25 workers working a 10 hours shift on a site employing 500 operatives.
- Recovery of re-usable materials (on one project of approximate value £200,000).
- Reduction in supplier journey times by delivering to the LCCC rather than going direct to the site, of an average of 2 hours per journey.
- Reduction of the number of construction vehicles entering the City of London and delivering to the sites being served by the LCCC, of 68%.
- Reduction of CO_2 emissions from reduction of vehicle movements of 75%.

Materials for site being loaded onto an LPG-fuelled delivery vehicle

Construction goods, excluding steel frames, aggregates and major plant, are delivered to the LCCC in relative bulk. From there, materials are called off by the various trade contractors and formed into work packs for immediate use on site, following a just-in-time approach.

Goods are checked on arrival at the centre for quality and condition, to ensure any problems are highlighted at an early stage.

The centre does not store goods in the conventional sense, with an aim of a turnaround time of 10–15 days.

Case Study 3: Consolidated Loads: The Answer to Space- and Access-constrained Sites (*Continued*)

Materials are safely stored

Materials are consolidated, which means that multiple part-loads are combined into single deliveries. This process maximises the efficiency of distribution vehicles and reduces the number of vehicles delivering into a congested environment.

The site productivity benefits from having a steady supply of materials delivered right to the point of use and keeping the skilled workforce at their work stations, doing what they do best.

Site housekeeping issues (quality, health and safety, waste and dirt generation) are greatly enhanced by the arrival and on-site storage of only those materials intended for immediate incorporation. At the end of the shifts, unused materials and packaging can be returned to the centre for recycling or reuse.

With its mission to deliver materials to site in the safest and most efficient manner, in active partnership with trade contractors and project managers, the LCCC has significantly benefited the various projects it services and contributed greatly to the achievement of the programme certainty demanded by the clients.

Several clients now view materials consolidation as an added insurance in the delivery of their projects and openly recognise that leaving individual trade contractors to fend for themselves is no longer the way forward.

Equally important, materials consolidation has a positive impact on relations with a site's neighbours, with the restricted flow of vehicle movements and associated emissions in any given location and time.

A recent study by BSRIA concluded that the LCCC methodology can deliver clearly defined and well-organised material benefits to projects and can do it consistently on time.

In September 2007, the LCCC moved from Bermondsey, SE16 to Silvertown, E16. The reasons for this included:

- larger warehouse space (6,000 sqm)
- better links with the City of London, Stratford and Docklands
- located on an excluded route on the London Lorry Control Scheme (London Lorry Ban).

All delivery vehicles used by the LCCC are LEZ compliant.

Partners Worked With

- Bovis Lend Lease and Stanhope: close working relationship initially as stakeholders and latterly as clients
- Transport for London (initial seed funders). LCCC provided data relating to construction traffic to assist Transport for London in formulating the London Freight Plan
- Metropolitan Police CVEU as a pioneer member of the Freight Operators Recognition Scheme
- working closely with principal contractors Skanska, Structuretone
- individual trade contractors, who appreciated the benefits to their business through taking part in the initial pilot scheme, are using the LCCC for other projects, including: Hall & Kay, Levolux, LA Construction, Spacedecks, Gartners, Mivan and Mag Hansen.

Main Benefits

- Goods are consolidated so that multiple part-loads are combined into single shipments.
- Substantial reduction in overall vehicle numbers delivering to a site.
- Goods are delivered not just to a site entrance but also to specified locations as close as is practicable to the workface, by materials handling operatives.
- Availability of specialist operatives who use an extensive range of vehicles and mechanical handling equipment necessary to complete distributions efficiently and without damage to materials.
- Overall coordination of distributions (to avoid clashes).
- 'Walking' of intended access routes, arranging road closures, lifting plans, ensuring that order is created in the distribution process.
- Trade contractors are left free to concentrate on their core tasks, without worrying about the coordination of and supply of goods to site.
- Trade contractors are not diverted away from production to assist with material handling.

Case Study 3: Consolidated Loads: The Answer to Space- and Access-constrained Sites (*Continued*)

Images of the Silvertown LCCC

Front and reverse side of a flyer produced to raise awareness of the LCCC in Silvertown (2007)

Case Study 3: Consolidated Loads: The Answer to Space- and Access-constrained Sites (*Continued*)

European Supply Chain Excellence Award Winner 2006: Environmental Improvement

Building Exchange (BEX) Winner 2007: Best Project Collaboration

Building Award Winner 2007: WRAP-sponsored Sustainable Construction

Green Apple Gold Award Winner 2007: Environmental Best Practice

Case Study 4: Come Up and See Me Some Time: The Value of Occupational Health

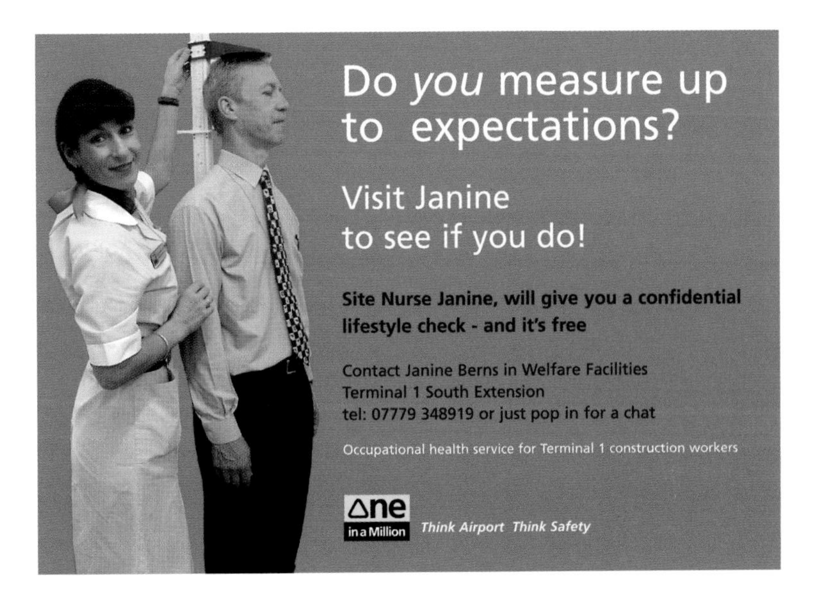

The importance of providing appropriate occupational health facilities for the construction workforce has been described in detail in Chapter 7. Construction workers often work long hours and spend a considerable amount of time travelling to and from site. Their travel and work pattern (and macho psyche) usually prevent them from having medical check-ups or seeking medical advice from their general practitioner when required.

This case study demonstrates that providing high-quality occupational health facilities on site brings significant benefits to the workforce, employer and society. In addition to providing enhanced first-aid facilities, national healthy lifestyle campaigns can be aligned with the occupational health services provided on site and focused on ailments that are particular to the majority male workforce (e.g. testicular cancer awareness, high cholesterol, skin cancer).

This case study also shows that a dedicated occupational health facility on site will significantly help reduce injuries, sick leave and identify potentially life-threatening illnesses that would otherwise go undetected until too late.

Case Study 4: Come Up and See Me Some Time: The Value of Occupational Health

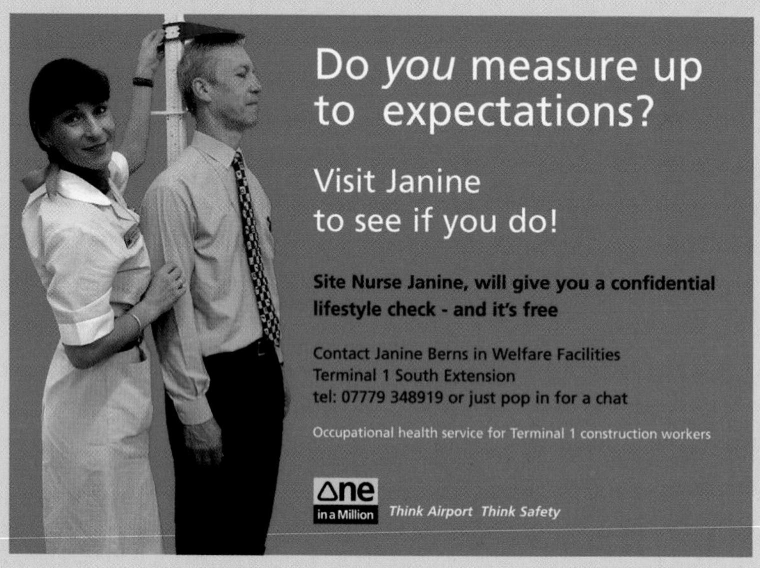

'One in a Million' poster used to promote the occupational health service at Heathrow airport

The provision of an on-site occupational health service has a few vital underlying elements:

- primarily aimed at reducing the accident frequency rate
- a client-driven innovation
- carefully performance-managed: impact is constantly measured against agreed criteria
- builds on the 'respect for people' initiative
- involves a full-time nurse on site.

Aims and Objectives

To promote the benefits of a healthier work environment, leading to measurable benefits, including:

- a reduction in the amount of lost time to construction through ill health
- a reduction in the number of accidents, minor and reportable
- the promotion of good health through training and education.

Benefits

- a confidential service available to all trade contractors on site
- alerting workers to possible health conditions that may otherwise go untreated
- supporting individuals living away from home, without access to a GP

- links with health and safety campaigns
- provision of on-site first aid; the ability to deal with minor injuries avoids the necessity to visit A&E when minor injuries are treatable on site, resulting in a reduction in productive hours lost
- multiple awareness campaigns: healthy eating, hand protection, testicular and prostate cancer checks, etc.
- the client clearly demonstrates that they value the workforce and, as a result, worker turnover is measurably reduced, particularly in the final phases of a project.

Most notable of all, where occupational health services have been provided, there have been a number of recorded cases of early detection of potentially fatal illnesses, enabling rapid pre-emptive treatment. We believe it is highly unlikely that these conditions would have been detected were it not for the readily available lifestyle MOT checks on offer.

From April 2002 to September 2002, Wilson James was able to collect data from the occupational health service it provided to BAA's Terminal 1 South Extension project at Heathrow Airport. Over a six-month period, the benefits demonstrated (strict programme of KPIs; source: BAA), included:

• Hours saved: treatment of minor site injuries	**504**
• Hours saved: treatment of non-site injuries	**566**
• Hours saved: workers from other projects using service	**96**
Total hours	**1166**
• Number of injury visits to service	**109**
• Number of medical visits (not site-related)	**145**
• Number of lifestyle checks	**135**
Total visits	**389**

Cost savings at a notional £35 per hour £40,810

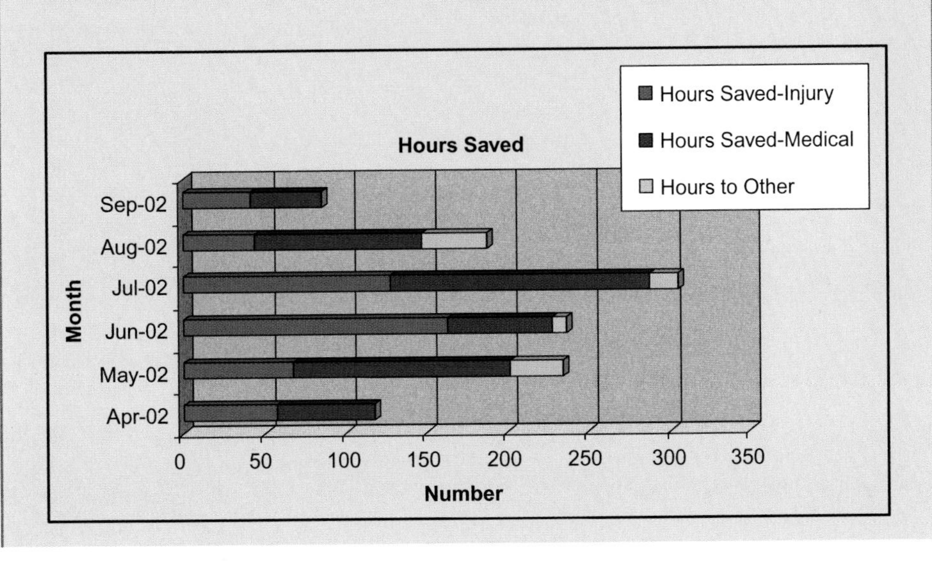

Section 3

Case Study 4: Come Up and See Me Some Time: The Value of Occupational Health (*Continued*)

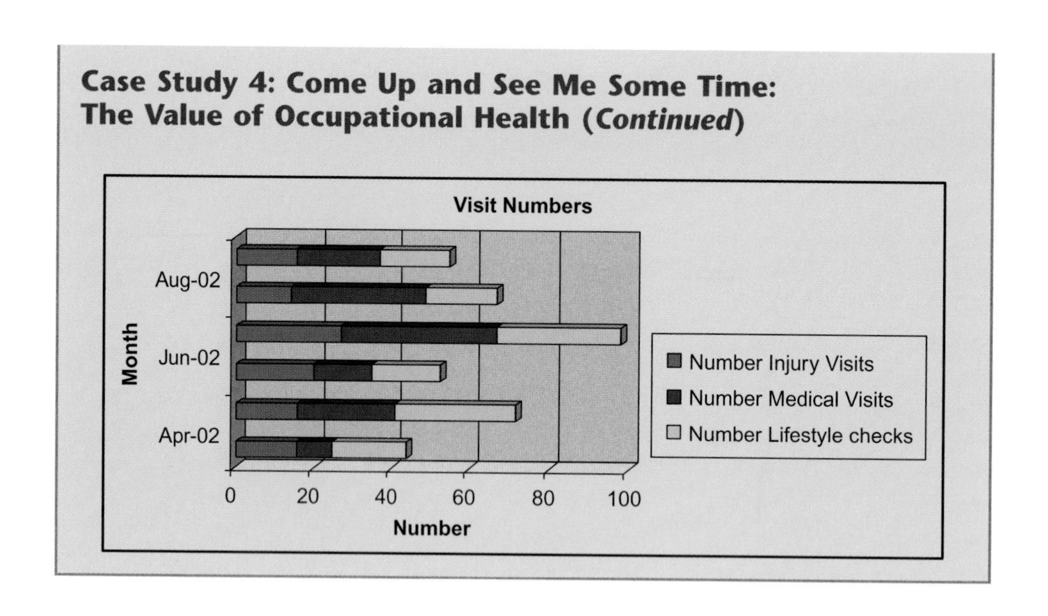

Case Study 5: Terminal 5: Delivering Europe's Largest Construction Project Just in Time

One hundred and sixty multi-skilled logistics personnel were employed at the prestigious Heathrow Airport Terminal 5 project. The project included a main terminal building, 60 aircraft stands, air traffic control tower, 4,000-space multi-storey car park and 600-bed hotel. Despite the enormity of this project, the site itself was very constrained and presented many logistical challenges.

By implementing a robust logistics strategy, significant cost savings were realised and productivity levels increased by around 25%.

Space for materials storage was limited and the logistics strategy had to manage 200 HGV deliveries a day during the peak of operations. At its peak, 3,000 construction personnel were on site, requiring high-quality office accommodation and welfare facilities.

The T5 project used a dedicated logistics centre at Colnbrook to control the flow of materials and ensure that they were delivered at the right time.

Case Study 5: Terminal 5: Delivering Europe's Largest Construction Project Just in Time

Heathrow Terminal 5

Implementing a robust logistics strategy at Terminal 5 realised significant savings. The increased reliability and efficiency in the use of materials led to an increase in productivity levels from a traditional build of 55–60% to 80–85%, with associated cost savings in the region of 2.5%.

Comprising the main terminal building, two satellite buildings, 60 aircraft stands, air traffic control tower, 4,000-space multi-storey car park, 600-bedroom hotel, all served by a new spur road from the M25, this £4.3 billion project was widely hailed as the new benchmark for UK construction.

Despite being equivalent in size to London's Hyde Park, the T5 site itself was very heavily constrained and presented many logistical challenges. To the north and south of the site were two of the world's most heavily used runways.

With 36 work areas on site, there was little or no room to store materials for any length of time. Project teams were required to plan their requirements for materials up to six weeks in advance.

Immediately after the initial planning stages of the project, BAA turned its attention to the site set-up and logistics requirements to service the undertaking.

Key facts

- Build programme stared in late 2003 and was completed March 2008.
- 3,000 construction personnel on site at peak.
- 200 HGV deliveries a day at peak.
- The logistics strategy is an integral part of the overall build programme and is recognised by all construction stakeholders at Terminal 5.

The Wilson James team was responsible for a variety of activities from erecting hoardings and partitions to providing the temporary accommodation, security, fire alarm and other systems. The next major step was to develop the materials delivery strategy.

The focal point for all construction material deliveries was the Colnbrook Logistics Centre.

Each of the various projects and contractors had access to a software booking system – Project Flow – a tool that has been structured especially for the construction industry.

Project Flow collates the team's demands and allocates a booking time into the Colnbrook Logistics Centre. It also identifies supplier, product, pallet weights and sizes for each delivery. These materials were then delivered either just before or on the day they were required.

Delivery vehicles being organised at the Colnbrook Logistics Centre

Each afternoon a report was issued to security and the booking-in clerk for deliveries the following day.

Consolidated loads were called off by the project using the same software for cross referencing and the load was delivered to the work front to a pre-arranged time.

Case Study 5: Terminal 5: Delivering Europe's Largest Construction Project Just in Time (*Continued*)

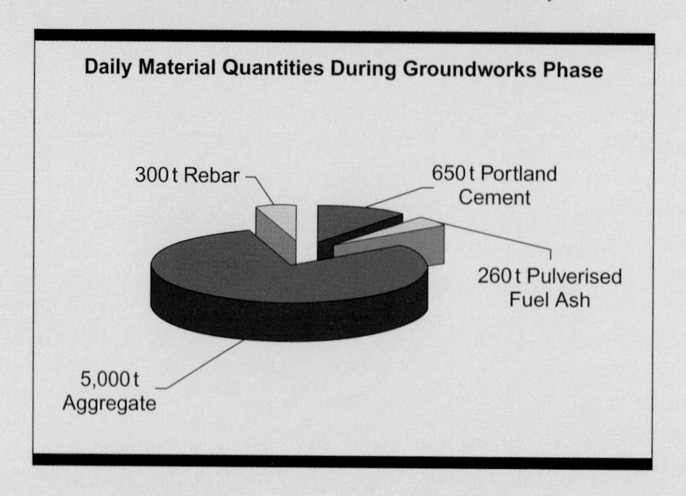

This process eliminated the need for extensive lay-down areas for materials, and the increased reliability and efficiency in the use of materials led to an increase in productivity levels. Cost savings were reported to be in the region of 2.5%.

During the initial phases of construction, T5 required a high volume of concrete and aggregate. Subsequently, the variety of materials used, such as steel, glass and the building services modules, increased significantly while the volume of materials declined.

An in-depth data-capture exercise was undertaken to establish stakeholders' requirements. This involved working with designers, framework contractors and subcontractors to establish individual site labour and management histograms, material types and volumes in line with planned productivity.

Once collated, this data was profiled to identify the peak activity, to enable accurate programming of subsequent resource levels.

- temporary hoarding
- multi-craneage use
- unloading gantries
- 15 hoist
- 22 forklift trucks
- consolidation centre

- 40 material unloading areas
- material delivery scheduling tool
- material delivery verification system
- waste removal
- multi-satellite welfare stations
- temporary electrical supply

- temporary access stairs
- weather reporting
- air quality monitoring
- edge protection
- water removal
- lifting exclusion zones

- temporary plumbing supply
- temporary signage
- fire safety
- evacuation procedures
- access control procedures
- 160 multi-skilled logistics personnel

Key principles identified for the logistic strategy

Artist's impression of Terminal 5 at night

Case Study 6: Security Services at Tate Modern and Tate Britain

Wilson James implemented a successful robust logistics strategy during the construction phase of Tate Modern and this successful relationship continued and developed into providing combined security, reception and logistics support when the Tate Modern opened as an Art Gallery in 2000. The success of this partnership led to Wilson James being awarded 'Best Partnering Initiative' at the Security Excellence Awards in 2005.

Security has been described in detail in Chapter 8 and is, unfortunately, considered a 'grudge expense' by many contractors: a service that is not accorded sufficient importance and is often undervalued. This case study demonstrates that in addition to providing the more obvious deterrent functions of security a holistic approach involving patrolling Rangers working in collaboration with Out Reach workers and the Metropolitan Police provides significant additional benefits to the community.

Case Study 6: Security Services at Tate Modern and Tate Britain

Tate Modern, International and Modern Art Gallery

Wilson James has worked in partnership with Tate Museums since 1997. The relationship has evolved from a logistics service contract during the construction of Tate Modern into a combined security, reception and logistics support service from 2000, when Tate Modern opened. Our service offering was further extended in 2005 when Wilson James was awarded the security contract for Tate Britain and Tate Stores.

Fully integrated construction logistics service, including:

- management of security systems
- control room monitoring
- access control and patrolling services
- reception/cloakroom services
- art handling/loading bay management.

Our relationship with Tate is not only based on our professional conduct but also on being able to:

- show an appetite for change and accommodating the customer's regularly changing needs

- offer an extremely diverse range of services, demonstrating Wilson James' considerable versatility.

This long-term partnership was recognised when Wilson James was awarded Best Partnering Initiative at the Security Excellence Awards in 2005, in recognition of its efforts in building a long-term mutually successful relationship with Tate.

In April 2006 Wilson James was appointed to provide a patrolling Ranger service for Bankside, one of around 30 'Improvement Districts' nationwide.

The Rangers cover an area in south-east London extending from Blackfriars Bridge in the west to London Bridge in the east and south to Southwark Street and a small part of Borough High Street. The team of four Rangers has a wide range of duties, including:

- looking after the local environment, reporting things such as safety hazards, graffiti, fly posting and ensuring that these are dealt with quickly
- directing and overseeing the cleaning contracts, e.g. jet washing and street cleaning
- meeting with businesses and traders in the area creating formal and informal networks to address concerns about safety and security
- keeping a watchful eye on areas around business premises and residential areas
- providing advice on crime prevention and safety issues
- promoting Bankside as a great place to do business, work, visit and live
- greeting and assisting visitors, giving them directions if required and escorting them to places of interest, also providing advice on local restaurants, transport and business destinations
- working closely with other agencies such as Out Reach workers, community wardens and the Metropolitan Police to deal with homeless people, beggars and illegal sellers.

Bankside Rangers on Patrol

Case Study 6: Security Services at Tate Modern and Tate Britain (*Continued*)

Dear Allison

I thought that I would drop you a line to say how impressed we are with the innovative and enthusiastic way that the Rangers are going about their duties. We have in a short time moved the service level up significantly from our previous arrangement and are now beginning to provide the sort of service that we envisaged a Business Improvement District being able to provide.

We are particularly pleased with the initiatives that are being taken and the assistance that is being provided to businesses, visitors and residents as well as the positive way that the Rangers are working with KGB, the cleaning contractor.

Of particular mention is the response to a problem with rough sleeping in the basement area of 52 & 54 Southwark Street over the last few days. Not only did the Rangers respond rapidly to a call for assistance from the businesses occupying the buildings, they alerted the Councils Street Population team, took advice from them on how to handle the situation, had the sleepers moved on (into hostel accommodation), arranged for the cleaning contractor to attend and had the site cleared and cleaned. All in short order. A model response to what could have become a serious and unpleasant incident.

Chris Bateman
Operations Manager
Better Bankside

Case Study 7: Refurbishment at Unilever House: Delivering Sustainability

The provision of welfare facilities has been described in Chapter 9, where it was shown that the standards of site catering and other important welfare facilities has evolved and improved greatly over the years. The canteen food standards on some construction sites can now be compared favourably with good restaurants, where the selection of nutritious, healthy and tasty food is provided by trained and qualified professional caterers – but at a much more reasonable price. Whilst the ubiquitous bacon roll or egg and chips remain favourites with the construction worker, there are sites today where they are able to get grilled sea bass and noodles or specially prepared foods that comply with religious or ethnic requirements. Catering in construction is becoming more sophisticated.

Welfare facilities also include the provision of changing rooms and lockers. This case study shows how a leisure-centre-style serviced cloakroom provided award-winning facilities and the approval of an appreciative workforce. Workers who have a choice of projects or contractors to work for are certainly going to prefer to work on a site that provides innovative, high-quality welfare facilities that enable them to look after their health and belongings better.

Case Study 7: Refurbishment at Unilever House: Delivering Sustainability

Facade of the Unilever House project

The refurbishment of Unilever House was presented with the Best Site Facilities Award in 2006 (*Building* magazine Health and Safety Awards). This award was achieved by the forward-thinking and innovative approach of Stanhope, Bovis Lend Lease, logistics provider Wilson James and Transport for London, who co-funded the London Construction Consolidation Centre at Bermondsey.

October 2004 saw the start of the three-year redevelopment of the Unilever House, six-storey, 400,000 square foot project located near Blackfriars Bridge in the City of London.

From the outset this project was designated by the Stanhope and Bovis Lend Lease Alliance as a demonstration project for materials logistics and waste reduction.

Key facts

- £200,000 of unused materials returned from the project instead of being disposed.
- Consolidated day work packs of materials delivered direct to specialist work forces.
- Just-in-time material delivery schedule.
- Waste materials were taken to the Construction Consolidation Centre for processing.
- 76% of waste materials were recycled.
- Reduced traffic flow in the surrounding area and increased productivity on site.
- Over 6,000 tonnes of steel and 17,000 m^3 of concrete were removed and recycled.

The Unilever House project team shared the aspiration of delivering the most sustainable project possible. Three of the team's key targets were to:

1. ensure key material deliveries to site were consolidated in south-east London
2. source construction materials locally wherever possible
3. reuse and recycle 95% of materials and equipment.

Award-winning welfare facilities at Unilever House

To facilitate these aims, the project was one of the first to benefit from the use of the London Construction Consolidation Centre (LCCC) at Bermondsey, where all materials for the project (except aggregates, structural steel, cladding and major plant) were initially delivered.

The use of the centre enabled the main barrier – a restricted delivery point in close proximity of the Crowne Plaza Hotel – to be overcome by ensuring deliveries were kept to a minimum, closely coordinated and timed.

Case Study 7: Refurbishment at Unilever House: Delivering Sustainability (*Continued*)

The logistics team conducted on-site toolbox talks to train contractors in the relevant logistics procedures. Information sessions for all site management were undertaken to ensure understanding of the LCCC methodology. A trade contractor information pack was issued to all parties. It contained the written logistics strategy and procedures.

Trade contractors were required to order their work pack call-offs and each order was recorded by the site logistics manager and forwarded to the LCCC for picking and delivery.

At the LCCC, materials were repacked into day work packs and forwarded to the site. Upon receipt at the site, the day work packs were delivered directly to the individual work faces, thus reducing traffic flow and congestion in the surrounding areas and increasing productivity on site. Individual contractor supply chains were established to focus on providing a robust just-in-time materials delivery schedule.

In all, some 13,200 pallets (or equivalents) were handled by the LCCC over a two-year construction period. At project completion the extent of unused materials returned to the centre rather than ending up in waste skips amounted to 38 full 26 tonne lorry loads, calculated at a value of £200,000.

Waste generation was recorded on a monthly basis and in total the project generated 3,384 tonnes at an average of 140 tonnes per month, of which a minimum of 76% was reused or recycled. In addition, 90% of all delivered pallets were returned to the LCCC for collection by the individual suppliers.

Award-winning canteen at Unilever House

Conclusion: The Argument for Change

Thanks to the wonders of modern technology, it would, at least in theory, be perfectly possible to create the software for construction to become a regimented and painless process. Picture the scene: the architect is sat at her computer.

She is working with a design program that allows her to drag and drop in standard components, some that can be pre-assembled, others that will have to be adjusted to fit the bespoke requirements of the design. As she works, the software automatically creates and updates a bill of quantities. Everything down to the number of nuts and bolts needed is automatically counted and adjusted as the design develops. This software is so intelligent that it knows the manufacturing lead times for all the required components. In fact, it is set up so that the architect can see how long it will take to produce, transport and install every element of her design. As soon as the design is finished, the core information for the logistics plan is available.

All the construction manager would need to do is sit down with an experienced logistician to identify any potential pitfalls and the entire requirements for the project could be planned and coordinated from the start. Technologically, this would be possible – as soon as a can of beans is swiped through the till of a supermarket, the information is sent back to the supplier in real time. This technology is now so advanced that we take it for granted that tins of beans will be on the shelves every time we go shopping.

Despite that, the likelihood of this system getting implemented in construction seems, at the time of writing, about as likely as Accrington Stanley FC winning football's Champions League. Given that, for a number of complex reasons, the space age construction logistics solution is not achievable, the great question is: 'What can we realistically do to improve things?'

As Mossman (2008) says, culture eats change for lunch. Although the top end of the industry is now embracing more professional solutions, the majority of projects still trust their luck to ad hoc logistics. It's taken twenty years for this first book on construction logistics to appear, and in the meantime the industry has remained resolutely underwhelmed by the growing body of case study evidence which points to the massive savings and efficiencies available. The issue of initial cost, and the contractual arrangements to cover them, could be overcome with a little application. What is much harder to change is the need to fire-fight – another issue recognised by Mossman.

However, as Peter Rogers states in the Foreword to this book, once a critical mass of firms has adopted the kind of professional logistics discussed in this book, this will become the normal way of working. It is the hope of the authors that this book will give lecturers assistance in introducing these ideas to tomorrow's industry leaders, and that the next generation will be ready to embrace the challenge of creating positive change through efficient logistical solutions for construction.

In the meantime, anyone interested in logistics can adapt the ideas in this book for almost any size or shape of project. Indeed, many of the issues discussed will already be very familiar to any experienced construction manager. Still, that construction manager may find this a useful text to give to his or her junior, since young managers often arrive on site knowing least about the mundane practicalities which they will initially be put in charge of – construction logistics has rarely featured in construction courses at the time of writing.

Why is this? The authors would suggest that this might be to do with the industry's complex relationship with its own body of knowledge. Historically, everyone on site deferred to the architect. It is well recognised that the tradition of deference stems from the days when the architect was a highly schooled middle-class professional who gave orders to the 'simple builder'. The authors know a lot of 'simple builders'. Many of them have a kind of genius about the vision and energy which they apply to the complex challenges of the projects they steward, but they tend to be incredibly self-deprecating. It must be remembered that, whereas architects and engineers have been chartered professionals for nearly 200 years, the Chartered Institute of Building only gained this status in 1980, when the political climate became particularly favourable to recognition for people in business. Prior to this time, there was generally a sharp distinction between professionalism and commerce. Those who worked for a profit were seen as socially inferior to the professionals, who worked for a fee.

In order to withstand this endemic experience of social inferiority, the builders have, over generations, built up a number of defences. For a start, they tend to have an uneasy relationship with academia. It has historically been customary to minimise the importance of mere book-learning, often whilst craving it. Now, construction management degrees contain the theoretical content to confer equal status with the other professions, but old habits die hard. Construction managers still profess themselves to be 'just' simple builders, thus covering themselves in interdisciplinary meetings from any inadvertent faux pas, whilst relying on a tacit knowledge of how things should actually be done in practice to maintain an advantage over their colleagues. This is their knowledge base. It has not generally become the provenance of the textbook. In the complex logic of this defensive inverted snobbery, albeit a snobbery of a most complex kind that relates solely to the intellect, it is the mundane which becomes most esoteric. The practical little tricks for doing things which the other professions aren't generally exposed to are often the builder's great unspoken pride. Builders keep their secrets about how things should really be done.

Perhaps this is why construction logistics – which represents much of the most mundane activity on site – is the only part of the entire construction process that is almost never designed. We design the building and the components and the landscape. We design the interior and the furniture. Logistics, however, is almost never designed-in, but usually grudgingly bolted on afterwards. Part of the builder's mystique, which has surrounded the profession since the days of the master masons, is the things they come to know through experience about how these mundane activities should be tackled. If there is any merit to this theory, then the cultural elements resisting change are deep indeed.

The traditional reliance on tacit knowledge means that lessons are rarely captured or shared. There is a certain romanticism to this reliance on private strategies to tackle complex projects, but it is increasingly inadequate to meet the demands of today's projects within the environment of today's contractual arrangements, with their anarchic dispersal of risk.

We began this book with a comparison with the military, and have mentioned their dictum that 'No plan survives first contact with enemy.' Similarly, no construction project survives first contact with the client. Despite this, formal programmes are often written as if set in stone. Meaningless process documents and tacit knowledge coexist, but rarely work together.

The army, by contrast, started to formalise its doctrine about 20 years ago. This doctrinal approach is not prescriptive or dogmatic, but it gives everyone involved a common framework within which to create solutions to particular challenges. With the demise of set-piece battles and the continuing modern war against insurgency, spelling out the basic approach has become more important than ever. It's about giving people the skill, the authority and the responsibility to adapt to changing circumstances, sometimes in very difficult conditions. The analogy with construction is obvious. The construction professional's personal armoury of skill and experience is as important as ever, but within a complex environment it could be put to far better use if everyone was working with a common understanding of the principles.

In Chapter 1, we saw how Bülow believed that a scientific approach to military manoeuvres could obviate the need for brilliance in the field of battle, saying that, 'the sphere of military genius will at last be so narrowed, that a man of talents will no longer be willing to devote himself to this ungrateful trade' (Bülow in Gat 2001). History proved him wrong, however. The man (or woman) of talents is as needed in the army as ever.

The same applies to construction. Everyone who works in the industry knows and loves the thrill of solving unique problems and meeting the deadline. A more orderly approach would not, as some might fear, create an industry too boring for the man or woman of talent; it would rather create an environment where that talent could shine and be used to greatest effect. If the authors had an aspiration for this book, it would be the hope that it might inspire the rising generation of constructors to look again at logistics, and realise that this can be as sexy and exciting as construction itself. Hidden beneath the mundane, and driven by increasing corporate responsibility obligations and more demanding societal expectations, is the power to create better conditions, improve productivity and prevent fatalities and injuries. That is as worthy a calling for the man or woman of talent as any other across the spectrum of construction professions.

Construction professionals have no further need for the traditional mixture of affected ignorance and arcane understandings. They stand as equals with the architects and engineers, shaping our skyline and our society with their skill and ingenuity. There is a sense in which they are the creators of our national identity, as well as of our homes, offices and landmarks. They are the leaders in an endeavour that is often, quite literally, to build someone's dream, and make their vision into reality. To do this, the author's believe that they need and deserve the best logistical support available, as a reflection of the pride which everyone involved in this remarkable profession is entitled to feel about the breadth of their talents and achievements. When cultural change comes, may it bring both better logistics and a better understanding of why construction is such an important and remarkable industry.

References

Gat A (2001) *A History of Military Thought: From the Enlightenment to the Cold War*. Oxford University Press, Oxford.

Mossman, A (2008) 'More than Materials: Managing what's needed to create value in construction.' Paper presented at the 2nd European Conference on Construction Logistics – ECCL, Dortmund, May 2008.

Index